FLORIDA
NATURAL SPRINGS

FLORIDA
NATURAL SPRINGS

• A History •

HOLLY SPRINKLE

THE
History
PRESS

Published by The History Press
Charleston, SC
www.historypress.com

Copyright © 2024 by Holly Sprinkle
All rights reserved

First published 2024

Manufactured in the United States

ISBN 9781467156974

Library of Congress Control Number: 2024938212

For my favorite springhunters, William, Michael, Daniel and David.

Special thanks to my Writing Squad and Dave O'Gorman for teaching me to trust the pixies.

CONTENTS

INTRODUCTION

In North Central Florida, a favorite pastime is "springhunting," or exploring freshwater springs. The spring water swirls in a spectrum of greens and blues, with a crystal clarity seen nowhere else in the world. Although the water temperature varies from spring to spring, most are seventy-two degrees Fahrenheit. Each spring has its own characteristics, such as the flow of water, recreational activities available, water depth and color and the aquatic and plant life in and around the water.

Humans have lived on the banks of these springs for centuries. In ages long past, the temperature and landscape of Florida were different than they are today. In some locations, giant versions of animals such as mammoths, sloths and armadillos called the springs home. The endangered West Indian Manatee stills flees the cold seawater during the winter months and uses the warmer spring water as a temporary home. The springs always have been, and remain, an important part of the ecology of Florida.

Life in North Central Florida has centered on the springs for a long time, a backdrop to the riveting stories that have become the legends of these waters and the communities around them. Legends sometimes arise from sensations shared by people, such as the vortex at Ginnie Springs, or a long-forgotten curse. Other times, a spring legend is based on a case of mistaken identity, like legends about mermaids that were most likely manatees. In some instances, however, the legend is based on a person who becomes larger than life and leaves a legacy that becomes an unforgettable story.

Springhunting is getting to know the springs and the legends that capture a moment in time on the banks of these enchanting waters. It's enjoying the water and the experience of visiting a special spot. Knowing the stories that surround these springs enhances springhunting. Each spring has its own story and its own set of characteristics that make it unique.

Most of us have tasted a spring. Spring water is captured and bottled by large companies that provide bottled water. The bottled water at your local gas station flowed below Florida for centuries before bubbling up to the surface. Currently, 2.3 billion gallons a day are pumped out of the Floridan aquifer, creating an imbalance. The water is drained faster than it can be replenished. The reduced flow can lead to ecological changes that affect the plants and animals that live in and around the springs and to collapses in the geological structures that are underwater. These collapses can cause serious situations such as sinkholes.

Protecting the springs is the best path forward to ensuring new stories and legends will continue to unfold on the banks of the waters that have changed and influenced the world.

This book will help you better understand the springs and their stories and what they mean to all of us. Happy springhunting!

Chapter 1

ICHETUCKNEE SPRINGS

Canoe Port, Kidnapping and a Lover's Lagoon

Ichetucknee Springs is a create-your-own-adventure, awe-inspiring national landmark. A hallmark of Florida's natural beauty, the pristine, clear, turquoise waters flow with a serenity that transcends time, inviting visitors to recognize their ancient stories. This sacred oasis is the backdrop of the stories born at this spring, rising from a time when the spring served as a focal point for the communities that lived along its banks. Over the centuries, this first-magnitude spring has flowed alongside a spectrum of communities.[*] The human stories centered on Ichetucknee Springs are about everything from a cruel, violent kidnapping to a destination for young love.

The name Ichetucknee is thought to come from a Native American word for "beaver pond." The present-day park is in Columbia County, Florida, an impressive collection of 2,669 acres with eight springs that feed into the one-of-a-kind Ichetucknee River. Those eight springs flowing into the river make the river water clear and easy to snorkel. The most popular activity at the Ichetucknee River is, by far, tubing. Due to the large amount of acreage that makes up the park, there are also several hiking trails where wildlife—including beaver, otter, gar, softshell turtle, wild turkey, wood duck and limpkin—can be spotted. Collectively, these springs produce two hundred million gallons of water per day. The spring water then feeds into the Ichetucknee River, which eventually joins the Santa Fe River and empties into the Gulf of Mexico.

---

[*] *First-magnitude* indicates the highest level of water flow.

Drone photo from above Ichetucknee Spring. *David Peaton.*

Though the park includes eight different springs, not all of them allow swimming or boat access. For example, swimming is not permitted in Cedar Headspring, named for the large cedar stumps that surround the spring. These massive stumps sit like tombstones for the huge oaks that were once part of the landscape surrounding the spring. The stumps are a reminder of the clearcutting in Ichetucknee's past and evidence that present-day forests around the springs would have looked much different as old-growth forests before the timber industry's influence.

Blue Hole is a favorite for scuba divers because of its depth. Its karst feature (a landscape feature created as limestone dissolves) is called the Jug due to both the clarity of the water and the shape of the area. Mission Spring is second in flow only to the Ichetucknee headspring itself. Devil's Eye has a unique feature in that the area has live oaks that sweep over the spring, twisting and turning, creating a unique visual effect. This spring is also known for bright blue reflections in the spring pool. Grassy Hole Spring is surrounded by wild rice. Mill Pond Spring used to power the gristmill when the Mill Pond settlement was an active community. Coffee Spring is home to the silt snail, which lives only in the spring. So far, this snail has not been found anywhere else in the world. This tiny snail is only a few millimeters in size and resembles a grain of sand. The snail, endangered and protected in the spring, is an example of the the natural springs' unique biodiversity. There are other unnamed springs that also feed into the Ichetucknee and provide an estimated 20 percent of the flow of the river.

Turtles at Ichetucknee Spring. *Laura Barone.*

Walking down a trail from the parking area and passing the bathrooms, a food truck and a picnic area leads to the swimming area for Ichetucknee Spring. For swimming, a staircase and a ramp make getting down to the spring easy. Entering the spring's swimming area is easy despite some rock formations at the shallow end of the spring, making water shoes a must.

There is also a ramp down to the swimming area making it more accessible. Although water levels are always changing, this spring is deep in the center and above the springhead. There is more deep water than shallow, so this spring can be a challenge for less experienced swimmers.

Snorkeling this spring is essential. The current can be strong above the springhead, but snorkeling directly above the springhead and watching the flow of water burst into the spring bed is unforgettable. Considering sixty-two million gallons of water flow through the spring each day, experiencing that natural power is awe-inspiring. The current from the springhead is strong enough that less experienced swimmers will need a lifejacket. Additionally, there are several different varieties of fish that visit the spring, including bream, bluegill, largemouth bass, alligator gar, mullet, catfish and several types of minnows.

For more wildlife spotting, consider tubing, kayaking or paddleboarding the Ichetucknee River in addition to the spring. Starting early in the morning is the best bet for seeing as much wildlife as possible, especially the abundant turtles that like to sun on rocks and fallen logs. More turtle species live together in the Ichetucknee area than anywhere else in the world. Gar, river grass and mullets are common to spot. Less commonly, manatees and alligators can also be spotted along the river. After tubing, there is a trolley that will take you back to the parking area, and there are also tubes available for rent, making this an easy adventure.

A better example of balancing human needs with environmental protection than other springs in the area, Ichetucknee benefits from measures being taken to protect the area. For example, tubing numbers are limited during peak season (Memorial Day through Labor Day) to protect the water quality, avoid erosion and limit damage to aquatic plants. During peak season, if tubing is part of the plan, the best idea is to arrive early or call ahead to see if the capacity for tubing has been reached. Tubing is worth the effort, though, as the river moves along at a relaxing rate (which can vary depending on recent weather).

Ichetucknee Spring is more than outdoor entertainment. For centuries, it has been the heart of the surrounding communities—for as long as people have lived in present-day Florida. The timelessness of this spring is a contrast to the ever-changing human story along its banks. Before it was even called Ichetucknee Spring, fossil records show that the area was inhabited by mastodons. Both bones and teeth have been found in the area, marking the spring as an important spot for both humans and wildlife even in prehistoric times.

Humans lived at Ichetucknee Spring for fourteen thousand years, at least. Only fragments of the stories of these early inhabitants are known, but the tools they used have been rediscovered at the spring and provide part of the picture. Paleo-Indians also sought out the springs for chert, a rock that is similar to flint. This material was used to make tools for hunting and other purposes. The earliest inhabitants of Ichetucknee Spring used stone weapons to hunt prehistoric beasts that are only seen in museums today. As hard as it is to imagine hunting a bison, horse or llama with a stone tool, it's even more incredible to imagine hunting the prehistoric, giant versions of sloth, tortoise or armadillo. Animals were attracted to the spring for its water, and hunters used this as an opportunity to hunt those that came there to drink. Hunters took advantage of the animals' vulnerable moment as they came for water and hunted in groups to swing success in their favor. Springs like Ichetucknee were life and death. The water sustained life in and around the spring, but for the animals hunted at their most vulnerable, drinking from a spring was also a risk, with ready human hunters waiting for that opportunity. Other than those suggested by their tools, stories of the everyday lives of these inhabitants have been lost.

More is known, however, about the inhabitants who came later, like the Timucua who lived at Ichetucknee Spring for centuries. Thanks to archaeology and a few accounts written after Europeans arrived, more details about their everyday life at Ichetucknee are known. The Timucua were not a single tribe but a collection of tribes that spoke a similar language. They grew crops but were also hunter-gathers. Although the Timucua presence around Ichetucknee Spring lasted for centuries, they began to live around the spring starting in 2000 BC. To put the long history of Ichetucknee Spring into context, the area was populated thousands of years before King Tut ruled Egypt. What would life have looked like along the banks of Ichetucknee Spring when the Timucua called it home?

Based on the descriptions from Europeans after contact, the Timucua were tall, often used face paint and had tattoos, which were a status symbol. The height difference between the Timucua and the Europeans was noticeable enough that rumors suggested there were races of giants living in what would later become Florida. Both men and women had long hair, which was cut when the chief died as a sign of respect and a reflection of mourning. They wore loincloths because they were in and out of the water and a loincloth dried quickly. Spanish moss and animal furs were also used for clothing, and in the winter, they added leggings for warmth.

Swimmers at Ichetucknee Spring. *Author's collection.*

As farmers, the Timucua grew pumpkins, cucumber, peas, maize and beans. Some fruits were also cultivated. Gatherers looked for berries and acorns. The spring and river allowed for hunting fish, turtles, oysters and deer. More dangerous animals, like alligators, were also hunted but not as often. Farming supplied enough food for half the year, but in the winter, the Timucua ventured farther into the forest to find other food sources.

Timucua villages similar to the one at Ichetucknee Spring would have consisted of huts built from palm fronds mixed with mud. The chief's hut was likely larger than the rest, and many Timucua towns had a community house where meetings and events were held. Villages also often had huts for storing food, and some had an additional hut for women to stay in right after childbirth.

Thanks to some of the artifacts found at Ichetucknee Spring, glimpses of Timucua life come into focus. Archaeological records show a very large collection of debris in one of the spring runs, similar to a garbage dump. As refuse such as shells was discarded, the pile grew over time into what archaeologists call a midden. Think about everything that can be learned about a person from going through their garbage, or how much information could be gleaned about a community and its habits from going through its garbage dump. The same is true for the ancient world. Not everything can be learned from a midden, but going through the garbage of this community at Ichetucknee provided shells, plant remains, animal bones, pottery and other artifacts. These items may not seem like they offer much information, but just the shells alone can provide information about what the community at Ichetucknee ate, where they traveled and who they may have traded with, based on their access to nonlocal items that are not found in the immediate vicinity of the spring. The midden and the records kept by Europeans after contact provide clues about what life would have looked like around the spring for the people who called it home.

Beyond a source of food and water, Ichetucknee Spring was even more important to the Timucua. Based on the contents of the midden, this spring served as a canoe port and a hub for their extensive trade network. At a time when rivers were aquatic highways and the main means of transportation, the spring was a convenient spot for canoes to pull off the river and trade their goods or load and unload items from a trade run. Things like pelts were valuable items to trade with other villages and, later, Europeans. The Timucua traded their goods for items like wheat, sugarcane, garlic, figs, melons and sweet potatoes.

The canoes were primarily used for fishing, but the trade network was also part of why they were built. They were carved out of the giant trees surrounding the spring, making canoes that were an average of eighteen feet long. The trunk of a tree was hollowed out, and these early canoes were often made large enough for fifteen to twenty warriors. Items that accumulated in the midden show how busy Ichetucknee would have been as a canoe port. Goods were coming in and out of the spring via canoe on

Replica canoe at the Silver River Museum. *Author's collection.*

a frequent and regular basis. The spring attracted animals that added to the food supply and provided pelts to trade. The water was used for drinking and agriculture, and as a final bonus, it facilitated the trade routes. The spring was a pathway to survival.

Going through the garbage of the Timucua people can provide glimpses into their everyday life, but it in no way unearths what they would have been like as people—and much of that information has also been lost. For example, the exact religious practices of the Timucua were kept secret, and only a few things are known about their belief system. The sun and moon were worshipped. They believed in omens. The chief served as a religious figure. Shamans were seen as powerful members of the tribe, and they used herbs for medical needs and provided spiritual guidance to the tribe. A shaman's other duties were varied and included delivering babies, finding new fishing or hunting spots, changing the weather and predicting the future.

Although some of their story has been lost to time, the final chapter of the Timucua that lived at Ichetucknee Spring is documented. Hernando de Soto visited the Timucua at their town of Aguacaleyquen (or Caliquen) in 1539. Arriving at Ichetucknee Spring, Hernando de Soto used a strategy of fear, violence and emotional manipulation to cross Native American land. His strategy was to kidnap a member of the chief's family and hold that person hostage until he crossed the land safely. De Soto offered no collateral to the Timucua or the other nearby tribes. Despite the language and culture barrier, de Soto asked each tribe he passed on his journey to trust that he would safely return their loved one. Considering that the tribe had no reason to trust de Soto and that his first introduction involved using threats and violence through kidnapping and capturing Timucua and other Native Americans to use as slaves, it would have been hard for them to imagine that the kidnapped person would actually be returned safe and sound.

Although this cruel strategy had worked in the past, it didn't work the way de Soto intended at the village near Ichetucknee Spring. During his visit in 1539, de Soto kidnapped the chief's daughter, as he had planned to do and had successfully done in other villages. Per de Soto's usual cruel routine, the chief's daughter would be held until de Soto's entire party crossed the Timucua land. Only then would the hostage be released. The chief would have had no way of knowing if his daughter would actually be returned, and the tribe was afraid of de Soto and his men. De Soto had created an impossible situation that escalated quickly. The chief had to decide whether or not to take the risk of de Soto keeping his daughter.

Observing the approach of the Spanish, the Timucua had already fled their village at Ichetucknee Spring ahead of their arrival. Once de Soto and company realized that the Timucua were hiding and fleeing, most of the Spanish went on ahead to continue on their planned journey. Violence could have been avoided all together.

When Hernando de Soto arrived at the spring, he captured twenty Timucua. The captives were added to the many Native Americans he had already kidnapped and forced into slavery. The chief's daughter was one of these captives. As any father would, the chief decided not to wait to see if de Soto would return his daughter after passing through their land. The chief may have considered that many of his warriors had also been captured and there was no way to ensure his daughter's safe return. Instead of complying and taking that chance, the chief and the remaining warriors followed de Soto, hoping to rescue her and the other prisoners. The chief's rescue attempt was ultimately unsuccessful, and he was also captured by de Soto. The conflict continued to escalate into a battle, and eventually, the Timucua warriors were chased into two ponds. Since they were familiar with the area, some were able to escape, but most did not. De Soto's kidnapping plot ended with many of the remaining Timucua captured and enslaved.

Despite this violent attack and catastrophic loss of life, the remaining Timucua continued to make their home at Ichetucknee Spring. The battle would not be the last time the Spanish visited Ichetucknee. Nearly seventy years later, the Spanish returned and built a mission at Ichetucknee Springs close to Fig Spring. A series of missions had already been built across Florida, but this one was designed specifically as a mission to the Timucua. The mission, called San Martin de Timucua, included a church, priest's quarters and a graveyard. The point of the mission, however, was not to coexist with the Timucua. Instead, San Martin de Timucua was meant to be a way to convert and control the Native population while providing a buffer against the English. The missions also became a way for the Spanish to more easily take the Timucua as slaves, forcing them into hard labor.

The impact to the Timucua population after Spanish contact was devastating. Before European contact, there were an estimated two hundred thousand Timucua living in Florida. Their civilization thrived across their thirty-eight chiefdoms, with their presence in Florida established for centuries. Part of what made the Timucua susceptible to European disease is that they did not keep livestock, so they had no immunity to the diseases people contract when living close to domestic animals. Disease, slavery and violence continued to destroy the Timucua population and their way of life.

In a brave last stand in 1650, the Timucua rebelled against the Spanish, led by Chief Lúcas Menéndez of San Martin, who put out a call to other chiefs nearby for warriors to gather and fight for their freedom. There were casualties on both sides as the rebellion began. The Timucua and their allies destroyed the mission, permanently. The Timucua left their villages and cornfields behind and waited for retribution. The rebellion was, however, unsuccessful, the warriors killed or captured.

The remnants of the Timucua were rounded up and merged into small villages outside of St. Augustine with other tribes and other Native Americans. The Timucua would never again reside at Ichetucknee Spring. Even the last survivors continued to disappear due to disease and being sold into slavery. The few who remained were eventually sent to Cuba in 1763 or became part of the Creek and Seminole tribes. The Timucua's centuries-long existence at Ichetucknee Spring and other parts of Florida is now a distant memory. Although pieces of their lives and culture remain, much of their way of life will never be known.

The human story at Ichetucknee Spring continued to change. Land grants were given out as a reward to some servicemen who fought in the Seminole War, providing a population boost to areas around Fort White. The cedars and pines that blanketed the area were clear-cut to make way for plantations. The water in the springs and river provided an opportunity to grow cotton, corn, sugarcane and oranges. A large plantation was later built at Mill Pond, and a still was built that predates the infamous Rum Island still.

In addition to the plantation, a sawmill was built, and timber from the forests that had surrounded the spring for generations was also used to build a general store, homes, a smithy and a gristmill along the banks of Mill Pond Spring. The end of the Civil War attracted displaced soldiers and survivors to nearby Fort White, with some of these settlers also joining the community at Mill Pond.

With plentiful resources, the community grew, and Ichetucknee was once again a home, now for this newest community of people. In 1878, the town of "Ichatucknee" got a post office. As it was for the people who came before them, the spring water and river water were the lifeblood of the community. Fort White reached a peak of two thousand people. Ichetucknee Spring was once again a place to swim, bathe, hunt, fish and worship, while the area also provided resources such as phosphate, citrus and cotton.

By 1841, there was a small settlement of twenty-one people including eight slaves that lived at Cedar Hammock, northeast of the Ichetucknee headspring. During this period, an army surgeon named Jacob Motte described Ichetucknee Spring:

The very air seemed concentrated in coolness, a grassy slope of the most rich and velvet green extended to the margin of a translucent and placid spring, whereon was faithfully reflected the green foliage that thickened over it; and in its transparent water might be clearly discerned the tiniest object at the bottom, clothed in the blent hues of the o'er arching sky; the babbling of the stream, and faint rustlings of the foliage as the breeze passed gently over the impending shrubbery, were the only sounds heard in this sweetest of sylvan solitudes. Various kinds of fruit trees glowing with blossoms were bright in loveliness around us.

Motte's words capture the tranquility of the spring. By the early 1900s, railroads were running from Lake City to the Ichetucknee area. More visitors began to come to Ichetucknee for recreational reasons, with the reputation of the tranquil water expanding to attract people from farther and farther away.

But 1890 brought yet another shift in the evolving Ichetucknee Spring area when the Dutton Phosphate Company started a mining operation at the spring, signaling the mining era of Ichetucknee Springs. Early mining in the area for phosphate was a simple process in the early 1900s. Wheelbarrows were used to carry the phosphate, and later, boilers, pumps and steam shovels were added. This type of mining caused less damage to the land.

The spring remained part of the lives of the local farmers as well, as people near the spring would farm during the day or use the water for irrigation and fish in the evenings. Watermelons floated in the water to keep them cool. It was a central spot for courting, eating, swimming and having fun.

Fifty years later, the Loncala Phosphate Company purchased the land around Ichetucknee Spring, and it would own the land for twenty years. Modern mining techniques damaged the environment significantly more than when low-tech methods were used. During the mining years, the mining company tended to ignore the swimmers who would sneak onto its land to swim. At first, it was just a few people. In the 1920s, the popularity of the spring rose, and it became a hot spot for everything from fish fries to baptisms.

People near bathhouse, 1920. *State Archives of Florida, Florida Memory.*

Ichetucknee Spring has an undeniable spiritual quality, and to see the spring is to feel its sacredness. Taking advantage of the spiritual quality of the spring, Ichetucknee became a hub for baptisms. Elm Baptist Church is one of the churches that frequently performed baptisms at Ichetucknee Spring. The recreational aspect of Ichetucknee also continued to emerge, and in the 1920s, locals built a bathhouse. The spring was so central to the local community that businesses in Columbia County closed early certain days of week. During these early closures, community members would meet for picnics and social and political gatherings and to simply take a break from the Florida heat.

Simpler pleasures were also part of Ichetucknee's past during this time. Residents recall eating boiled peanuts while enjoying the scenery or getting a Christmas tree from the area where the mining used to happen. Old mining pits provided changing areas before there was a bathhouse. Sometimes stews would be made at the spring and eaten all day, shared communally, similarly to a stone soup (where each person brings an ingredient to add to the pot of soup as it cooks). Locals would bring squirrels they caught near the spring or feral pigs and add that to the stew. The spring provided entertainment, and the community came together to provide food. Moonshine, called white lightning by longtime Columbia County residents, would also be brought and shared in jugs at the spring. Field and farm workers were bribed with swim time in the spring if they worked harder or faster. The spring water was also taken in barrels as drinking water.

Next, Ichetucknee Spring became a spot for romance, a regular lover's lagoon. Among teens and young adults, the spring became an easy spot to meet because of its seclusion. As the center of the community's social life, Ichetucknee Spring had a reputation for late-night lovers' tryst. Long before online dating, Ichetucknee Spring was where locals went to meet a romantic partner. Ichetucknee Spring was Cupid's pool, with locals remembering getting engaged or courting while sitting on the rocks at the spring.

The spring's popularity had some downsides, though. One was the cars that ended up in the water. Longtime residents of Columbia County described a teenage girl accidentally running her car into the spring. The 1936 Ford ended up getting pulled out with mules. Another resident remembers a couple that ended up in the water in their vehicle: "A couple parked close to the banks of the Head Spring and began necking. Soon the passionate session ended when the car rolled into the spring." An unplanned dunk into Ichetucknee's seventy-two-degree water could not have been how the couple envisioned ending that evening.

Although the spring was important, it was also not protected yet. During this time, dynamite fishing became part of Ichetucknee's story. Used as a way to catch a lot of fish at once, dynamite fishing was exactly what it sounds like. Much to the detriment of both the fish and the fisherman, a quarter stick of dynamite was lit and thrown into the spring. When the dynamite exploded, the shockwave from the explosion either killed the fish immediately or stunned them, making them easy to capture. Although there were locals who had lost one or both arms using dynamite to fish, it remained an accepted practice for a time. Longtime visitor to Ichetucknee Spring Thelma Lowe described it this way: "They'd drop 1/4 stick, fish would go belly-up, and they'd scoop them in a basket." It was easy to tell which fish had been caught this way, as their backbones were usually shattered. The practice left large holes in some of the rock formations in the spring area. This type of fishing was not only dangerous for the fisherman but also damaging to the spring bed.

Blasts of dynamite for fishing and the chaos that must have followed vehicles slipping into the water did not impact the Ichetucknee's reputation. The seclusion, the romance of the scenery and the (mostly) quiet privacy the remote area provided continued to make Ichetucknee a hot spot for singles. Carolyn Cornman remembers this time at Ichetucknee Spring. She had a memorable romantic experience in the spring. A friend of hers arranged a blind date at the spring, a common practice in Carolyn's small hometown. When Carolyn got to the spring, she discovered her date was her cousin. Surprisingly, Carolyn described the date as the best one of her life: "I used to be shy around boys, and on that date, I didn't have to worry about a thing." The popularity of the spring as a spot to meet dates meant risking meeting a family member!

Historically, the banks of the Ichetucknee would have been no stranger to loincloths, but it's Blue Hole that escalates this family of springs from romantic to wild. During this period, Blue Hole got the reputation for skinny dippers. Taking the path to Blue Hole was synonymous with naked swims. Other springs in the area were spots for nude bathing as well, but Blue Hole was the popular place to go. Averill Fielding remembers taking a wagon hitched to a mule up to Ichetucknee to go swimming on a typical day, saying, "You didn't need a bathing suit if you swam at Blue Hole." New Year's Day is the official day to skinny dip at Ichetucknee, where at first light on the first day of the year, people congregate to swim and usher in the new year, nude and in nature.

Swimmers at Ichetuckee Spring,
1920–26. *R.W. Blacklock.*

Road technology and construction paved the way to the next shift in the human story at Ichetucknee. With the development and improvement of roads leading to the spring, Ichetucknee became a favorite spot for students from the University of Florida. It was an adventure to leave the "big city" of Gainesville and wander through the woods to find the spring. Like the location of a secret party, the spring's location spread by word of mouth. Students enjoyed the spring in a different way than that of Ichetucknee's past fans and visitors.

In the 1950s, University of Florida students discovered the spring in larger numbers, and it quickly became a hot spot for large, unplanned parties. Students would grab a tube or swim and drink college party amounts of alcohol. Eventually, Ichetucknee became popular not only for nude swimming and but also for drugs, particularly hallucinogenic and psychedelic drugs. The deep blue hues of the spring water already look like something from another world, making the spring a memorable backdrop for using mind-altering substances. In 1966, University of Florida students were at the height of their LSD experimentation years. Later, students enjoyed a popular strain of cannabis called "Gainesville Green." Shrooms, also known as magic mushrooms, grow naturally in this region of Florida. The springs offered an enticing amount of privacy and adventure for students compared to their everyday campus life. The spring was a serene setting for experimentation.

Not everyone was pleased by the students' interest in partying at the springs. By the 1960s, said John Hill, a resident at the time, "The situation was out of control, with thousands of tubers and parties around each weekend." For a time, each weekend outdid the one before it. Ichetucknee Spring's popularity continued to intensify. The spring was the place to be. Where else could you go as a University of Florida student and swim naked in crystal-

Women on floats at the spring. *State Archives of Florida, Florida Memory.*

clear water, partaking in a variety of substances with other students while floating in a tube? The raging parties at Ichetucknee rose to infamy—so much so that by 1968, Ichetucknee Spring would get state attention as local residents continued to voice concerns.

Part of what became a problem for the larger community was the aftermath of the students' parties, particularly litter. Come Sunday, the students would return to Gainesville, leaving an amount of litter now seen only after a home Gator game. Piles of trash, drug paraphernalia and popped river tubes were left behind. The majority of the garbage, though, was beer bottles and cans. John Hill said of the issue, "The trash and vandalism problem was overwhelming." Community members who grew up visiting the spring did not like the direction things were going. They were tired of the litter and the party atmosphere.

Another community member, Mr. Porter, remembers this period. "UF students were lying around naked and doing drugs," he said. "About four or five truckloads of beer bottles were pulled out from under the bridge." There was also some damage done, such as concrete being thrown into the spring.

By the late 1960s, families were no longer coming to the spring because of the "drugs, drunks and trash." The students and their activities at the spring had overtaken Ichetucknee. The spring was still owned by the mining company, but it overlooked the students' activities, at least for a while.

At this point, Governor Claude Kirk got involved. Randy Robinson of the Florida Highway Patrol was asked by the governor to "clean up that den of iniquity at the Ichetucknee River caused by drunken college students." Robinson focused on the US 27 bridge, and he had many "unpleasant" encounters with the students. In the first weekend after the mandate, Robinson made thirty arrests for drunk and disorderly conduct and nudity. The governor made the parties stop, and students moved on to other options. Another era ended for Ichetucknee Spring.

Ichetucknee Spring was not just a den of iniquity during this era, though. Beginning in 1958, students and community members began making demands that the state take over control of the spring, as it was still owned by the mining company. The Save Ichetucknee Springs advocacy group promoted governmental oversight of the area in order to save it from private development.

Visitors unload their supplies from the float at the spring. *State Archives of Florida, Florida Memory.*

Complaints continued to mount about the damage being done to the spring. The mining company had to constantly deal with vandalism, trash, debris and complaints from local residents. Although the mining company was nonchalant at first, eventually it became tired of dealing with the issues at Ichetucknee. When concerns were raised about the damage being done to the spring and the surrounding natural areas due to more invasive mining practices, the company decided to sell the land to the state in 1970. The future of this spring changed once again, although it took years of cleanup work to undo the damage done.

The state worked on cleaning up and protecting the spring to get the area to what it is today. The varied history of the area, however, has led to some legends that remain due to the mystical vibe Ichetucknee has always had. The combination of the remoteness and the quietness of the area has led to a legend from hikers exploring the woods around the spring that there are invisible entities in the forest. Visitors to the spring report hearing someone walking, even though the forest appears empty. Some hikers in the area report hearing steps that are as loud as stomping behind them. When they turn around, nothing is there.

Swimmers, especially on emptier days, report hearing whispering while swimming in the spring. When they rush out of the spring to see where the sounds are coming from, the woods around the spring appear empty. Some swimmers report a more sinister voice that whispers to them, encouraging them to cause harm to themselves. Once again, there is no indication of where the whispers originate. Ichetucknee Spring has stood as a witness to so many stories that it certainly must have a few secrets left to be discovered.

What does the future look like for Ichetucknee Spring? It still attracts quite a crowd, mostly for tubing in the summer months. The park has set limits on occupancy that help reduce damage to the spring and river. Ongoing threats to the spring include agriculture runoff and changes to the water flow from water bottling companies. A common argument in favor of allowing water harvesting by bottling companies is that it does not impact the flow of the spring boil. Leonard Bundy, a former park ranger at Ichetucknee Spring, recounts how much the spring volume has changed. He most vividly remembers the summer of 1944, when the main springhead had such a volume that you could hear the spring boil from the parking lot. Other accounts from this period report that it was hard to swim across the spring because the pressure from the springhead was so intense. Take a moment to listen at the parking lot when visiting Ichetucknee Spring. There is no sound of the spring boil's water.

A helpful way to think about agricultural runoff is to picture a bathtub. Substances put into the groundwater seep into the aquifer, just like water draining from a bathtub. These pollutants then show up in other places that connect to the Floridan aquifer in springs such as Ichetucknee.

Ichetucknee Spring stands as not only a natural wonder but also a living testament to the intertwined narratives of humanity. Ichetucknee Spring has been the core of the community around it for a long time, but local communities are not the ones that will determine what happens to the spring next. Industries like timber and mining have taken a toll in the past; the current corporate threats are the agricultural and bottled water industries. Al Burt, a reporter for the *Gainesville Sun*, put it this way in a 2003 article: "Springs nourish clues to our natural past, and they encourage us to recognize that what we have left is too precious to squander on hucksters who never sipped from a spring while keeping an eye on a crawfish at the bottom, or on a snake hanging off a tree limb overhead." The spring's future must be decided by its local community. With any luck, the natural beauty of the spring and the stories that originate from the timeless journey of the water will inspire future generations to continue modern conservation efforts to protect the spring and allow it to serve as a setting for the next chapter of Ichetucknee Spring stories.

Chapter 2

MANATEE SPRING

The White King's Party, an Important Alligator and a Sublime Spring

Manatee Spring is a flagship spring in Levy County, Florida. In 1968, Manatee Spring became the first freshwater spring in Florida to become a state park. Manatee Spring is one of those spots that, like Ichetucknee Spring, feels ancient. Above and below the blue-green water, Manatee Spring is a visual reminder that the springs are from another age. Though many have been lost, a few stories of what life would have looked like around this spring have survived. Different communities called Manatee Spring home long before the headspring got its current name. After one European botanist's visit to the spring, it became an international sensation that permanently changed the literary world and helped define the style and voice of American literature. Manatee Spring inspired the world, and the impact of that moment in time can still be seen today.

Manatee Spring feels immersive and ancient. That feeling is partly due to the aquatic life that has thrived in these waters for ages: largemouth bass, speckled perch, catfish, bream and gar still enjoy Manatee's waters. This first-magnitude spring releases one hundred million gallons a day. Manatee Spring is surrounded by swamps and hardwood wetlands and is situated right off the Suwannee River. The park covers 2,443 acres. The wooded areas around the spring have sinkhole ponds with unique features. There are four main cavern openings in the spring: Manatee, Catfish Hotel, Sue Sink and Friedman Sink. One of the sinkhole ponds has a unique cave ninety feet below the surface that connects to Catfish Hotel. Catfish Hotel is named for the abundant catfish that divers typically see in the spring.

Drone photo of Manatee Spring. *David Peaton.*

Manatee Spring Run joining the Suwannee River. *David Peaton.*

The four areas allow for different activities. Scuba diving is allowed in Catfish Hotel, but Sue Sink is an emergency exit only. Friedman Sink is available as an entrance to the cave system but only to certified drivers. Manatee Spring offers recreational options: hiking, biking, kayaking, canoeing, scuba and snorkeling.

As the name suggests, this spring is also a good viewing area to spot manatees in the winter months. The best chance of seeing a manatee is in the fall or winter, when the manatees seek the warm spring water as they travel inland from the Gulf of Mexico. The spring is a safe haven for the manatees, which cannot survive cold water. Some of the manatees pick the spring as their spot to calve. The presence of manatees, however, is not a guarantee either. When manatees are present in the spring, no swimming is allowed. Black vultures also winter at the spring in impressive numbers.

The Manatee Spring cave system is one of the longest explored cave systems in the world, with twenty thousand feet (about the length of sixty city blocks) of caves mapped. Although the spring does have frequent scuba and cave divers, it can be a difficult dive due to the strong flow and low

Ichetucknee Spring. *David Peaton.*

visibility in certain conditions. Still, the structures under the water are a ready reminder that Manatee Spring has had an abundant life under the water since well before the first divers entered the spring. Within the cave system are fossils of ancient animals that visited the spring long before humans lived along its banks.

The Manatee Spring area was inhabited by humans for centuries because of the fresh water and food sources that the spring provided. The proximity of the spring to the river also made this spring a great spot for transportation, as the Suwannee River can be taken all the way out to the Gulf of Mexico. Artifacts found at the spring suggest that human populations have lived along the banks of the spring for at least nine thousand years.

A picnic area on a hill above the spring was once a thriving Native American village. The earliest known residents of Manatee Springs besides Paleo-Indians were the Timucua, who also settled at Ichetucknee. A park ranger described the location where the village once sat. "The whole picnic area was a Timucuan Indian village site," Ranger Maphis said. "They chose the site because it provided access to the Suwannee River for transportation. It also provided plenty of fresh water that was clean. In addition to that, it provided a food supply." The village likely had several thatched huts and a larger dwelling in the center. The spring would have made meeting the minimums for survival much easier.

The first time Manatee Spring became an international sensation was long before it attracted the dive community's interest. Much like the modern cave divers interested in exploration at Manatee Spring, William Bartram was a naturalist who had a deep passion for nature and was, understandably, captivated by Florida. He had been a caretaker for his family's garden for years, nurturing his interest in botany, but it wasn't until after his travels in what would become Florida, Georgia and the Carolinas that he became famous for his descriptions of nature in the areas he explored. Bartram traveled these areas and others well before the United States was a country. While exploring, Bartram wrote the late 1700s version of a travel blog. He included descriptions of the people and communities he met along the way and sketches and drawings of the plants and animals he encountered. Many of the plants, animals and Native peoples Bartram encountered would have been new to his European readers back home.

With Bartram's fame and the wide appeal of his book, Manatee Spring experienced the eighteenth-century version of going viral. Europe was captivated by the enchanting crystal waters Bartram described. The book was so popular that it was translated into several other languages, which made the

Above: Manatee Spring. *Author's collection.*

Opposite: Title page of William Bartram's book. *State Archives of Florida, Florida Memory.*

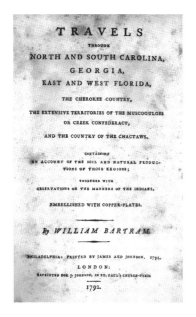

text even more of a sensation. The botanist who set out to prove that plants and animals of North American were noteworthy had succeeded. Europe wanted to know more about Florida, its creatures and the magic blue waters of the springs.

To describe Manatee Spring, Bartram called it the "incomparable nymphæum"— *nymphæum* meaning "idol to the gods of the springs." Bartram's communion with nature, his passion for the natural world, was infectious. His descriptions have a spiritual quality as he emphasizes the vastness and grandeur of Florida landscapes. In Bartram's book, nature is not passively observed and noted. Instead, nature is depicted as a force that excites wonder and awe. Although Bartram was scientific in his descriptions, he was also enraptured by the scenes he described. That enthrallment with nature made Bartram's work influential for a long time after its publication date. Nature began to be seen as a transformative force, a place to be enlightened.

Though the springs may have been news to the Europeans reading Bartram's book, for the Seminole and Creek tribes, they were a part of everyday life and were central to their survival, just like they were for the Timucua in centuries past. One entry in Bartram's travel record that captured the majesty of Florida waters described the first time he saw a spring, even before he encountered Manatee Spring. His guide (likely an enslaved person) had already explained to Bartram that the spring does not have any tributaries and does not originate from a lake. Another Native American Bartram met explained that the water originates from deep in the earth. Bartram was fascinated with the spring's water and also explored the Suwannee River, both of which he described as having an abundance of aquatic life. In Bartram's words:

It is amazing and almost incredible, what troops and bands of fish, and other watery inhabitants are now in sight, all peaceable; and in what variety of gay colours and forms, continually ascending and descending, roving and figuring amongst one another, yet every tribe associating separately.

Bartram's description of schools of plentiful fish swimming with one another in the waters as if each is variety is part of a tribe is a marked reminder of how much things have changed in terms of how plentiful aquatic life is now compared to Bartram's time. True to form, the scientific parts of his observations (such as fish counts) are paired with poetic descriptions of the same fish.

The poetic nature of Bartram's writing was also echoed in his descriptions of the water itself. When visiting a spring today, walking up to the water can be breathtaking if it's your first time seeing the clarity and colors of a freshwater spring. For Bartram, the same was true. He portrayed the impressive natural beauty of the springs.

> *We now ascended the crystal stream, the current swift, we entered the grand fountain, the expansive circular bason* [basin], *the source of which arises from under the bases of the high woodland hills, near half encircling it. The ebullition is astonishing, and continual, though its greatest force or fury intermits, regularly, for the space of thirty seconds of time: the waters appear of a lucid sea green colour, in some measure owing to the reflection of the leaves above: the ebullition is perpendicular upwards, from a vast ragged orifice through a bed of rocks, a great depth below the common surface of the bason, throwing up small particles or pieces of white shells, which subside with the waters at the moment of intermission, gently settling down round about the orifice, forming a vast funnel.*

Bartram's scientific training became important here. He was able to capture details such as the flow cycle of the water. His timing of the flow of Manatee Spring—every thirty seconds—was correct, proven so centuries later. Bartram carefully described the pressure and intensity of the spring flow and captured how the current from the springhead is strong enough to make canoeing in the water above the springhead difficult.

> *At those moments, when the waters rush upwards, the surface of the bason immediately over the orifice is greatly swollen or raised a considerable height; and then it is impossible to keep the boat or any other floating vessel over the fountain; but the ebullition quickly subsides; yet, before the surface becomes quite even, the fountain vomits up the waters again, and so on perpetually. The bason is generally circular, about fifty yards over; and the perpetual stream from it into the river is twelve or fifteen yards wide, and ten or twelve feet in depth.*

As important as Bartram's descriptions of the natural world were, he also gave a glimpse into Native American life. His observations, however, were made through a European lens. Bartram's journal entries from his time at Manatee Spring give clues about what life around Manatee Spring would have been like well after Europeans arrived in the area. However, they are just a glance and do not represent the entirety of what it would have been like to have been a Native American person living at the spring. By the time Bartram arrived at Manatee Spring in 1774, the Timucua had already lived at the spring for centuries and were long gone. The archaeological record and the ruins of their once-thriving village were the last remnants of their time at Manatee Spring.

Bartram rode along with a trader from St. Augustine who had been asked by the governor of East Florida to buy Seminole horses thought to be descendants of Spanish horses. These horses were desired by Europeans because they were known to be "the most beautiful and sprightly species of that noble creature, perhaps anywhere to be seen." Acquiring horses was Bartram's official reason for the trip, but his attention was focused on exploring and continuing to write and capture what he saw in his ever-growing "travel blog."

When Bartram visited Manatee Spring, although the Timucua village that once boasted at least thirty huts would have already been reduced to ruins, the Seminole had a settlement nearby. The Seminole were relative newcomers to the area. Their presence was so recent that their era in the Manatee Springs area would have been after Blackbeard was a pirate in the Carolinas. The Seminole grew their numbers by welcoming freed slaves and members of other tribes. Freed slaves were seen as equals. Their town was governed by a *mico*, or chief, named the White King. Bartram described the White King as follows:

> *The White King of Talahasochte is a middle-aged man, of moderate stature; and though of a lofty and majestic countenance and deportment, yet I am convinced this dignity, which really seems graceful, is not the effect of vain supercilious pride, for his smiling countenance and his cheerful familiarity bespeak magnanimity and benignity.*

When Bartram arrived at Talahasochte, the White King and several of the men from the town were out hunting. Bartram and company decided to wait for a few days, spending their days at Manatee Spring and surrounding areas and going back to town each evening. Bartram described thirty or so

Reflection in Manatee Spring run. *Author's collection.*

homes in the village, built "in the tradition of Cuscowilla" with a council house in the center of the homes.

The Seminole had an extensive and impressive trade network extending as far as Cuba and the Bahamas. Using large canoes that were built from the massive cypress trees that once towered around the spring and could hold twenty to thirty warriors, they traveled across the water trading their goods. The Seminole would have traded alligator, pelts and foodstuffs. Some of these canoes have been found and preserved thanks to the cold water of the spring.

At the time of Bartram's visit, a group of warriors had just returned from Cuba. The warriors brought back liquor, coffee, sugar and tobacco. To get these goods, the warriors had brought with them deerskins, furs, dried fish, honey and bear oil to trade.

While waiting for the White King to return from his hunting expedition, Bartram was invited along on a fishing trip, where he saw huge schools of fish. Long before overhunting, overfishing and pollution would diminish these numbers, Bartram's descriptions of these great schools of fish would be part of what attracted additional Europeans to come to Florida. Bartram and his companions were met with hospitality and warmth and even live music:

> Soon after joining our companions at camp, our neighbors, the prince and his associates paid us a visit. We treated them with the best fare we had, having till this time preserved some of our spirituous liquors. They left us with perfect cordiality and cheerfulness, wishing us a good repose, and retired to their own camp. Having a band of music with them, consisting of a drum, flutes, and a rattle gourd, they entertained us during the night with their music, vocal and instrumental.

Bartram lingered in this area to meet the White King, uncertain of how long it would be until he returned. For Bartram, waiting for a few days was worth meeting a man with the reputation that the White King had. A local plantation owner, Denis Rolle, who set out to build a utopia in Florida, met the White King years before William Bartram. In 1765, Rolle described the White King as follows:

> [I was] soon after sent for by the head Man of that Tribe, who bears the Name of the White King. I went immediately to his Hut, and found him, with six or seven other stout Indians, sitting on their Couches of Repose. The Chief enquired of the Reason of our Journey, and at first seemed to object to our proceeding farther; but after some Time, he seemed to say nothing further. Soon after there was served up some Venison dressed with Bear's Oil, and a Bowl of China-Briar-Root soup. They invited me up to a Dance, which they use on the Arrival of Strangers, and the whole Village

joined in it till about Eleven o'clock. The Chiefs came down likewise, and they seemed to be also in a very agreeable Humour.

When the White King and his party of warriors arrived back to their town, the villagers hosted a feast in honor of their guests. The White King and the warriors with him on the hunting trip had killed several bears, so bear was on the menu. When the hunting party arrived, a fire was lit in the public square. The royal standard was displayed, and drums were played to let everyone in town know there was going to be a feast. Bartram described the music that was played as "irresistible." It especially left an impression on him.

There is a languishing softness and melancholy air in the Indian convival songs, especially of the amorous class, irresistibly moving, attractive, and exquisitely pleasing, especially in these solitary recesses, when all nature is silent.

The menu was lavish as the White King went all out to feed his guests. The main dish was one of the bears that the warriors brought home from their hunting trip, but that was just the beginning of the feast. In addition to the bear, hot bread and honeyed water were served. After months on the road, this feast must have felt decadent and exciting. The meal was served in the banquet house, with the White King served first. Next the warriors were served and finally the guests, including Bartram. The rest of the town was served last. The food was served by slaves.

After the feast, the town switched to party mode. The townspeople returned to their cabins and huts for the evening, but that didn't mean the night was over. Calumets (ceremonial tobacco pipes) were shared after dinner. Black drink, a popular caffeinated tea made from steeping the leaves of a certain variety of holly, was served after the feast as well.

Although the black drink signaled the end of the feast, music continued to be played, and men and women danced in an afterparty that lasted throughout the rest of the night. Bartram was given supplies for his ongoing journey, including bear ribs, venison, fish, turkeys (also known as the "white man's dish"), hot corn cakes and a type of jelly. With a belly full of bear, after an evening spent frolicking, Bartram fell asleep. Around midnight, he was woken up by sounds of a struggle. After investigating, Bartram realized his companions were fighting an alligator near where he had been sleeping. Bartram described the run-in with the alligator:

Giving the alarm to the rest, they readily came to his assistance, for it was a rare piece of sport. Some took fire-brands and cast them at his head, whilst others formed javelins of saplins, pointed and hardened with fire; these they thrust down his throat into his bowels, which caused the monster to roar and bellow hideously, but his strength and fury were so great, that he easily wrenched or twisted them out of their hands, and wielding and brandishing them about, kept his enemies at distance for a time. Some were for putting an end to his life and sufferings with a rifle ball, but the majority thought this would too soon deprive them of the diversion and pleasure of exercising their various inventions of torture: they at length however grew tired, and agreed in one opinion, that he had suffered sufficiently; and put an end to his existence. This crocodile was about twelve feet in length: we supposed that he had been allured by the fishy scent of our birds and encouraged to undertake and pursue this hazardous adventure which cost him his life. This, with other instances already recited, may be sufficient to prove the intrepidity and subtilty of those voracious, formidable animals.

Bartram's companions tortured the alligator until they eventually ended its life. This wouldn't be Bartram's last encounter with an alligator: his descriptions of them include haunting images of alligators fighting with each other, feeding so loudly in a lake that the noise interrupted sleep, snapping at him while he walked down a path and following his canoe as he hunted for fish. One of the most disturbing images Bartam describes is a lake full of alligators, where he claimed that so many were feeding on thousands of fish that the lake was filled with streams of blood. He described the alligators as so plentiful that a person could walk across their backs to get to the other side of the water. It is this juxtaposition of beauty and terror that Bartram captured in his writing—an apt way to describe Florida.

In his book, Bartram included a crude drawing of alligators. Although his drawing looked nothing like an actual alligator, that seemingly important detail did not end up mattering. Bartram's readers were enthralled with the alligator. While earlier books suggested that the so-called New World had feeble animals, Bartram's narrative ushered in a shift in how Florida was seen by Europeans. Florida became not only exotic but also a place where both the power and majesty of nature could be experienced at once. Once his book was published, it created a boom in sales of Florida lands, inspiring Europeans to try their hand at taming and profiting from Florida spaces.

Drawing by William Bartram of two alligators in Florida, 1773. *State Archives of Florida, Florida Memory.*

It was also at Manatee Spring that Bartram saw a manatee skeleton, on the banks of the spring. The manatee was left there after the Native Americans hunted the animal the previous winter. Given that manatees can weigh thousands of pounds, the Native Americans likely harvested what they could and left the rest of the carcass on the spring shore. Bartram carefully measured parts of the manatee, which were included in his description. Bartram likened the bones of the manatee to ivory, which would have sparked even more interest in the creature.

> *Part of a skeleton of one* [manatee]*…lay upon the banks of the spring… The flesh of this creature is counted wholesome and pleasant food; the Indians call them by a name which signifies the big beaver. My companion, who was a trader in Talahasochte last winter, saw three of them at one time in this spring: they feed chiefly on aquatic grass and weeds.*

Manatees are no longer allowed to be hunted due to their status as an endangered species. Today, the primary threats to manatees are challenges with the availability of food, frequent severe blooms of red tide (an overgrowth of algae in the ocean) and physical injuries from boats. Much has been learned about manatees since Bartram's encounter, but the details he shared about the "big beaver" were accurate. His description of the

44

Bartram's drawing of an alligator.
William Bartram.

carcass of a manatee—which would have been another new creature for his audience at home—led to the spring being called Manatee Spring.

Nearly three hundred years before William Bartram saw and measured the manatee skeleton at Manatee Spring, Christopher Columbus wrote about the creatures—only he didn't recognize manatees as manatees. Instead, Columbus thought he was seeing mermaids. In a note in his journal from 1493, Columbus wrote: "Monday, 9th of December. They [sailors] saw three mermaids, which rose well out of the sea; but they are not so beautiful as they are said to be, for their faces had some masculine traits." What the crew thought were mermaids were actually manatees—perhaps because the sailors saw the manatees at a distance, when they were likely fatigued, leading to this case of mistaken identity. The legend that Florida has mermaids persists.

Bartram's descriptions were much more accurate than those of the sailors reporting mermaids. As important as *what* Bartram described was *how* he described the natural world he encountered in Florida. Bartram's was the earliest written description of the spring, and the language he included feels spiritual and romantic. Bartram's European audience was fascinated with the novelty of creatures like alligators and manatees (even if Bartram's drawings hardly looked like alligators).

Bartram's drawings of alligators and stories of encountering the reptile were the "clickbait" that got Europeans interested in his book, but it was his descriptions of Florida waters and the landscapes he walked through that left a lasting impression. Bartram's rapture with Manatee Spring was contagious. The emotional intensity in his descriptions of the natural world later influenced poets like Samuel Taylor Coleridge and William Wordsworth. The poetry of Coleridge and Wordsworth captured the idea of nature as sublime. This idea of the sublime—a pleasurable, intense feeling of awe when experiencing the beauty of nature—became a marked and important influence in literature. Transcendentalism takes this idea of the sublime one step further, suggesting that witnessing the sublime can lead to a spiritual experience. Bartram's writings became an influence on Coleridge's famous poem "Kubla Khan":

In Xanadu did Kubla Khan
A stately pleasure-dome decree:
Where Alph, the sacred river, ran
Through caverns measureless to man
Down to a sunless sea.
And from this chasm, with ceaseless turmoil seething,
As if this earth in fast thick pants were breathing,
A mighty fountain momently was forced.

Coleridge's description captured the beauty and power of water, as did the work of other Romantic writers. Bartram's descriptions of sacred waters like Manatee Spring were echoed by American transcendentalists who came later, such as Ralph Waldo Emerson and Henry David Thoreau, as well. The joy and power experienced in nature is seen when Bartram describes Manatee Spring: "This charming nymphaeum is the product of primitive nature, not to be imitated, much less equaled, by the united effort of human power and ingenuity!" Although much of Bartram's writing centers on the sublime, literary scholars like Mary Alice Kirkpatrick explain that his importance goes beyond influence alone: "Still more significant, William Bartram, in recounting his joyous, solitary communion with the natural world, emerges as one of the first writers to treat a theme that might be considered uniquely American." Bartram's travels in Florida and his stop at Manatee Spring helped to define the distinctive American voice. He was the first to capture a writing style that was later recognized as uniquely American and became part of the United States' identity. American literature often reflects nature as sublime or nature as a means to enlightenment, and Bartram had an influence in that development.

The Seminole presence in the Manatee Spring area would continue until the Seminole Wars, when their numbers decreased. After the Seminole Wars, the area was popular with timber farmers, who clear-cut the ancient forest and displaced the remaining Native American population in the spring area.

Today, Manatee Spring is best known for its cave system. Although unique, fascinating and beautiful, underwater cave exploring can also be dangerous. Across the state of Florida, three hundred cave divers have died exploring areas such as Manatee Spring. In some cases, divers get lost in the cave system because they get turned around, become disoriented, lose contact with the guideline or underestimate the strength of the current in the spring. The unsung heroes of these dives that end in tragedy are the rescue cave divers who search for divers who don't return from the spring when expected. The mystery and tragedy of Manatee Spring's cave system have led to some legends about the spring.

Manatee Spring is also known for having swift undercurrents in addition to the extensive underwater caves. The headspring is powerful and forceful. These strong water patterns along with the spooky stories of divers who have lost their lives in this spring have given rise to a few legends about this spring. Some locals claim that these strong, swirling waters are the spirits of divers who have perished trying to pull swimmers under the current. One woman claims to have felt an icy hand pulling her under the water. She was dragged deeper into the water, where she felt several icy hands pulling her downward until she hit her head on a rock. She was able to get away but stated she would never return to the spring again. Other accounts state that there are long plants that will wrap themselves around swimmers' feet and pull them under the current. This could be in part due to the long strands of river grass that used to grow in the spring.

As Manatee Spring has long been a spot for manatees to winter, one of the legends about the spring involves mermaids. Several extinct relatives of the manatees that currently frequent the springs had more of a fish-like tail. The infamous Christopher Columbus even sighted a manatee that he called a mermaid—more specifically, an ugly mermaid. Columbus's account is one of the earliest written records of a manatee being spotted and documented. In fact, manatees being mistaken for mermaids was common enough that the scientific name for these sea cows is *Sirenia*, a name reminiscent of myths about mermaids. There is also a Seminole legend about miniature people living at the bottom of a spring.

When visiting Manatee Springs, besides swimming in the beautiful, cold, clear water, make sure to take some time to enjoy other features of the park. A relaxing raised boardwalk takes you through the woods along the spring run all the way to a floating boat dock and a large covered gazebo where you can enjoy sights of nature, view the fish swimming in and out of the spring, bask in the amazing view of the Suwannee River or even do some fishing off the boat dock. Families can enjoy a small concession stand and use one of the many picnic tables that are around the main spring or near the playground. For those who enjoy hiking, several long hiking trails throughout the park take you through the woods, where you can view the amazing landscape. Primitive and RV camping is also available, but spots fill up quickly, so reservations are recommended.

Manatee Spring cannot be a hospitable ecosystem for mermaids or manatee populations while Florida's waters continue to struggle with red tide and other perils. Agricultural runoff has created an excess of nitrates in the water, which leads to excessive amounts of algae. The algae prevents other aquatic plants from growing, causing fewer fish to be present in the spring. The legendary swimmer-grabbing seagrasses can no longer grow in the spring run due to the changes in water quality. It's hard to imagine the spring in its present state even being able to support a village the size of the Timucuan village at the time it flourished. Bartram saw the spring at, perhaps, its peak beauty and clarity. His writing about Manatee Spring and other natural wonders changed literature and helped define the American voice—certainly, a spot and a story worth saving.

Chapter 3

TROY SPRING

Lost Cities of Troy, a Runaway and an Unforgettable Steamboat

Set back in an eighty-acre forest (about the area of a large shopping mall), Troy Spring is surrounded by sandhills and pines. Ninety-six million gallons of water a day flow through this first-magnitude spring in what is now Lafayette County. What makes Troy Spring unique among Florida's freshwater springs is its depth. This spring boasts an average depth of seventy feet at the springhead. That number may not have mattered when the Suwannee's only water traffic was canoes, but once steamships were part of the story, deep water was important. This state-owned park is home to two prehistoric archaeology sites and a historic Civil War–era shipwreck. Troy Spring is a deep, watery ship graveyard that tells the stories of a time when Floridians relied on the waterways for their way of life. The two cities of Troy, unfortunately, had a fate no better than that of the ship at the bottom of the spring.

Troy Spring is a popular spot for snorkelers and scuba divers. Recreational swimming is not as popular at this spring due to its lack of shallow water and limited entry options into the spring swimming area. Entry into the spring is currently limited to a long walkway that has limited areas for sunbathers to place their personal items and no real space to casually enjoy a leisurely swim in the spring. The stairwell going into the spring plunges the swimmer immediately into very deep, cold water, with almost no area for less experienced swimmers to stand and enjoy the water. At this spring, life jackets are a must for less experienced swimmers. For the adventurous, however, this spot offers different type of landscape than other nearby springs. Bring a

Boaters at Troy Spring. *Karl Holland.*

snorkel, some fins or a dive buddy and enjoy what is sometimes called the underwater Grand Canyon of Florida.

Like several of Florida's springs, the archaeological record at Troy Springs shows signs of human inhabitants beginning around twelve thousand years ago. The Paleo-Indian presence in the area was determined by the presence of stone tools such as blades and knives. After that early activity, the landscape of Florida went through significant changes due to rising water levels and a shrinking landmass. By 7500 BC, the area around Troy Springs had increased in population, with evidence of more specialized tools and a more settled lifestyle. By 2000 BC, pottery was in use in the cultures found in the Troy Spring area.

After European contact, life at Troy Spring changed permanently. Faced with disease, enslavement and violence, the remnants of the early tribes were forced to migrate to other areas and later joined the fragmented Native peoples who would later become the Seminole tribe. By the Second Seminole War, most of the Native population around Troy Spring was permanently gone.

The benefits of the spring appealed to Europeans as well. The Spanish were present in the Troy Spring area, but it was another group that gave

During the reconstruction days, this county fared like most others in the South where bands of deserters, Carpetbaggers and freed negros robbed, burned out and stoled away the stock of all the better families leaving them destitute. The court house, in fact the whole town of Troy was burned and destroyed during one of these raidscausing many of the records of the county to be destroyed. Later the town was rebuilt and so remained until the county seat was moved to Mayo in 1893. The first court house in Mayo was built in 1895.

Excerpt from the yellow notebook historical record of Lafayette County. *State Archives of Florida, Florida Memory.*

the spring its current name. To survive, several families decided to band together, and the first city of Troy was born. These families used the ancient trees around the spring to craft log homes. Some of these logs have been recovered from the spring; others still lie at the spring bottom.

The founding date of the city of Troy is not known, and even its exact location remains up for debate. In fact, everything about the city of Troy is debated by historians, including its location, who lived there and what eventually led to the town's demise. Much of what is known about the history of the area is taken from notes by an unknown author typed on a yellow notepad that was stored at the courthouse as an official record.

What is known, however, is that the city of Troy was somewhere along the spring's banks, but frequent flooding has made pinning down the exact location difficult. Most often, the location of Old Troy is set as being right on the banks of Troy Spring due to the writings of Mr. Melton, an early Floridian who wrote about the area and some of its townspeople.

The town's beginnings were humble. Even if the spring offered aquatic food and a water source, that did not mean life along the Suwannee River was easy. The Suwannee is a powerful river, so when floods happen, they come quickly, which would have meant people had little warning to leave Troy. The water also could take weeks to go back to a more normal level. Based on where Melton claimed the town was located, the town itself would not have flooded easily, but the trails leading back to larger towns and settlements would have also flooded easily. This would have left the town of Troy on its own to survive until the water level went back down.

Even with the location and the exact founding date of Troy unknown and up for debate, what is known is that by 1845, there was a functional post office. Eventually, Troy grew to include a county courthouse and other structures. The trees along the banks of Troy Spring were used to continue to make more buildings, and the town flourished.

The depth of the spring, which would have made living along Troy Spring challenging, was now an incredible asset. When the Suwannee River was a

major transportation route, steamboats were the lifeblood of the settlements along the river. The steamboats carried goods that the towns needed for survival. The boats were also a way for the residents of Troy to sell their goods to other spots along the river. The water depth made Troy Spring especially attractive to these steamboats, the saviors of the small settlements along the Suwannee.

At times, a single person shapes a chapter of a spring's legacy. For Troy Spring, that person was James Tucker. After his father died unexpectedly, James Tucker's mother married a man James could not tolerate. The young man came to despise his stepfather. As a means of survival after finding his living situation no longer tolerable, James ran away from home—right at the point that steamboats at Troy Spring became vital to everyday life. He took the first job he could find, and that was on a steamboat on the Suwannee. James Tucker learned what he could about life on a steamboat because this was a readily available job for a young runaway. Soon, he was finding his own way and making a life along the water.

James Tucker learned steamboating as a means to survive, but he saw an opportunity along the Suwannee. Tucker became an established boating entrepreneur in the Suwanee area. During this time, James Tucker was stashing away two important things: dollars for his own boat and knowledge

Opposite: Troy Spring. *Author's collection.*

Above: View of swimming area at Troy Spring. *Author's collection.*

of the Suwannee. Those dollars eventually added up, and Tucker was able to travel to Indiana to have his own steamboat built. After years spent learning the skills to steamboat, James Tucker was finally ready to captain his own ship.

The steamship *Madison* was born, named for Madison, Florida. Sometimes ships and boats become their own characters on the water, and that was certainly true of the steamship *Madison* along the banks of the Suwannee. The *Madison* was made and built for the Suwannee waters specifically. Captain James Tucker was clever enough to give special instructions to have the ship made a very shallow draft. This was a smart plan by Captain Tucker in that the Suwannee River has frequently changing

Portrait of Captain James Tucker. *State Archives of Florida, Florida Memory.*

water levels, a challenge faced by towns like Troy. By making the draft of the boat shallower than that of other steamboats, Tucker ensured that the steamship could navigate the river even when water levels were lower. The shallow draft on the *Madison* also meant that Captain Tucker could go to spots along the Suwannee that other ships could not reach without risking damage. Tucker's design for the *Madison* gave him an immediate competitive advantage. Towns along the Suwannee that didn't have the advantage of deep water, like Troy Spring, could still get supplies and buy and sell what they needed, regardless of the water level near them. The ship access meant that when other towns were cut off from other settlements due to flooding, Troy wasn't and could continue getting supplies.

With the ship build complete in 1854, the *Madison* began service on the Suwannee. The *Madison* was a point of excitement as she sailed up and down the river. According to a newspaper reporter who lived and wrote during these years, "[No vessel] has ever excited more interest than the steamboat Madison did to us scattered Crackers, along the Suwannee River in the days before the advent of railroads, or the beginning of the Confederate war." Carrying mail and serving as a floating trading post, the *Madison* became a staple of the towns and outposts along the Suwannee.

Tucker's design specifications, based on his years learning the Suwannee after running away from home, were a game changer. His ship made its way farther down the Suwannee River than any other craft could at the time.

A steamboat similar to the *Madison*. *George Baker.*

Tucker's route started in Cedar Key; then he would sail the *Madison* up as far as the river's height would allow. Captain Tucker liked to push the limit and push the *Madison* to go farther and farther down the Suwannee River. Due to the way he had the *Madison* designed, his ship could go farther up the river than his competitors', but Tucker wanted even more. He wanted to take the *Madison* to a spot where no one else was even trying to go and that everyone thought was impossible to reach. Given that this was during a time when Florida's rivers were the primary means of transportation for supplies, the *Madison* was an important part of daily life for those living along the Suwannee River, and Captain Tucker's vision was to keep connecting more settlements together.

The arrival of the *Madison* was exciting for the residents of Troy, as well as for those who lived in other towns along the river, as the steamship was a floating general store in the 1850s. The contents of the *Madison*'s cargo would change, adding another element of excitement to her arrival at each dock. Each week, the *Madison* started in Cedar Key and would stop on the shoals right above Troy Spring, two miles north of Troy. Produce was picked up from the homesteads that were isolated along the river, including venison, cowhide, deerskin, tallow, beeswax, produce, honey, chickens, eggs and hogs.

At each landing, Captain Tucker would blow the *Madison*'s impressive whistle. Another feature of her design: the *Madison*'s whistle could be heard from ten miles away. When the *Madison* was built, Tucker asked that it have the loudest whistle possible. Given how scattered some settlements along the Suwannee were, Captain Tucker wanted everyone along the river to know when he was approaching a dock. As he approached a landing, Tucker would begin sounding the whistle to let homesteaders know he was getting close, giving them time to make it to the water. By the time Tucker got the *Madison* docked, a crowd would be waiting at the dock, ready to make their trades or buy and sell cargo. He would stay docked at the landing for as long as there were people who wanted to trade, purchase or sell goods.

If the water level was high enough, Tucker would take the *Madison* as far as the popular White Springs. Tucker even pushed to have the Suwannee declared navigable, but opposition in the Florida legislature prevented that declaration from being made. The hesitation was because no boat had

Troy Spring meets the Suwannee River. *Author's collection.*

Swimmer at Troy Spring. *Author's collection.*

navigated down the river past Columbus. Captain Tucker saw this as a challenge. According to a newspaper reporter at the time, "When this state of affairs was reported to Capt. Tucker he swore that he'd be damned if he didn't put the Madison in White Springs if he had to run her up there on wheels." Tucker had a mix of determination and fearlessness that made him able to achieve things that others were afraid to do. The line between being cocky and an innovator was thin for Captain Tucker. He decided that the risk of damaging the *Madison* was worth taking to prove that the Suwannee was navigable farther down the river.

Captain Tucker waited until the Suwanee had a good deal of rain. As it is often prone to do in Florida's more watery seasons, the Suwannee flooded, with its waters reaching the forests along its banks. To prove his point, Tucker pushed the *Madison* up the river in these difficult conditions. He made it to White Spring, as he was determined to do, but the *Madison* took the damage of his determination. With the banks flooded, the water level was higher, making some aspects of the trip easier. But the flooding also meant there were more downed and hidden trees that could potentially damage the ship. As he kept pushing to White Spring, the trees ripped parts of the *Madison* off. The *Madison*'s smokestack and pilot were gone. The high water level granted him passage, but the flooded conditions and damage to the trees along the river made the tall smokestacks a target. They were caught on branches and the downed trees, which destroyed these parts of the ship. Captain Tucker knew what he was risking, and even as parts of the ship were ripped off, Tucker continued to White Spring and made it. Before repairs were even completed on the *Madison*, the Suwannee was declared navigable. Tucker had made his point to the Florida legislature: a win for Tucker, even though the *Madison* paid a high price.

The *Madison* and her voyages up and down the river, now with an expanded route, were a highlight of life in Florida during this period, and in many ways, this ship and others like it made life along the river possible. The ships not only brought supplies, but passengers also began to use the steamships as a way to explore Florida, and early tourism in the area became easier. The *Madison* turned out to be an even better investment than Tucker could have ever imagined. Always looking out for new opportunities, with the Suwannee River declared navigable, Captain Tucker bid on a mail contract to take mail up and down the river. Tucker's route started at the Gulf of Mexico, and the *Madison* made its way down the Suwannee twice a month carrying mail. This contract took Tucker from being successful to being very wealthy.

For the young teenager who turned to steamboats as a way to survive, that choice paid off, as Tucker received $30,000 a year for delivering the mail. His investment in the *Madison* was the best gamble Captain Tucker ever made. The way the *Madison* was able to bring supplies to so many settlements changed the Suwannee River Valley forever.

Tucker did not, however, get a happily-ever-after ending, and neither did the *Madison*. Conflict would come to Troy Spring once again, and the *Madison* served as a warship of sorts.

As the Civil War accelerated in 1861, the *Madison* took a commission as a supplier for the brand-new Confederate navy. The *Madison* was loaded with supplies that were later transferred to waiting trains. At first, life as a warship was not that much different for the *Madison*. Since his early life, when Tucker saw an opportunity, he grabbed it. Serving as a commissioned ship was that next opportunity.

Then the ship took on quite another purpose, and Tucker's luck started to shift. With their fates seemingly linked, the *Madison*'s journey took a huge turn. The steamship was fitted with guns, and the Confederates intended to use the ship as a privateer. The *Madison* served as a makeshift gunboat for two years. Instead of the savior of the settlements along the Suwannee, the *Madison* became a force to be reckoned with during the war.

The *Madison* was also used to investigate the four ships used by the Union in Key West to transport supplies. The steamship was enlisted to eventually take control of Union ships. Then, in 1863, James Tucker and his crew were asked to fight in Virginia. Tucker used his connections along the river and raised a company of men. If he was going, Tucker was going to go big and take as many able-bodied locals with him as possible. Tucker and company were then sent to Virginia, and Tucker's days along the Suwannee were over.

The residents of Troy and others living near Troy Spring asked Captain Tucker to use the ship one more time to take a load of corn down the river. Tucker gave his permission, and he and the *Madison* parted ways. Before he left, though, Captain Tucker gave one last set of instructions that sealed the *Madison*'s fate. A journalist described this critical turning point for the *Madison*:

> It was about September, 1863, if the writer remembers correctly, when the *Madison* was abandoned by Capt. Tucker….A number of citizens living near Troy wanted the boat to bring a load of corn up from Old Town and Capt. Tucker turned her over to them, told them to use her as long as they wished and then sink her for him in Old Troy Springs. The load of corn was duly brought to Troy, unloaded, and at 2 o'clock in the afternoon on a bright sunshiny day.

Tucker's order changed Troy Spring forever. After that last load of corn was delivered, the ship was scuttled (intentionally sunk) in the spring, where it still rests today. Troy Spring became the final resting place for the *Madison*, despite the ship being Tucker's most prized possession. Troy Spring was a good spot to sink a ship, as the spring's water depth once again came into play.

The plan, however, was never for the *Madison* to be in the spring permanently. Instead, Tucker thought that at the end of the war, the ship could be raised out of the spring water. The spring was deep and the water always cold, providing a place where the *Madison* could be kept safe in his absence. Tucker did not want the *Madison* to fall into enemy hands. He intended to hide the *Madison* in the depths of Troy Spring until he could return, resurrect her and continue their life together along the Suwannee.

Unfortunately for the *Madison*, that is not how things went. By the end of the war, every usable piece of the *Madison* that could be taken had been stolen. Her machinery and fittings were gone, and after a time, only the hull remained. Tucker was also gone for longer than he expected. When he left for Virginia, he thought the war was nearly over. Although Tucker would leave the Confederate army due to health problems, he would never resurrect what was left of the *Madison*. The macabre end to a steamship that once brought so much joy to the Suwannee River residents was described in a newspaper story:

> [They] *ran the Madison from Troy landing into the spring, pulled out her plugs and sat there and watched her till she rested on the bottom. During the war her boilers were removed, split lengthwise, carried to the sea coast and used in the manufacture of salt. Her smokestacks were cut up into convenient lengths and used by neighboring farmers as funnels for their sugar furnaces. The cabins were torn up and the lumber used by whomsoever wanted it, and when the war ended, all that remained of the Madison was her hull resting on the rocks under the crystal waters of Old Troy spring—and there it remains today.*

Parts of the *Madison* were scattered across the Suwannee River Valley for various purposes. Today, the hull still sits at the bottom of the spring and can be seen with snorkel or scuba equipment. The shipwreck site is the skeleton of what is left of the *Madison*.

For all practical purposes, the Civil War halted steamboat activity on the Suwannee River for a time. The first Federal naval raid in Florida was against Cedar Key, the Gulf Coast end of the Florida Railroad and a center for blockade running. A landing party from the USS *Hatteras* descended on the town on January 16, 1862, and destroyed the railroad depot and wharf, seven freight cars, the telegraph office, warehouses, three sloops, four schooners and a ferry barge. With the fall of Cedar Key and the resultant control of the nearby Suwannee by Federal blockaders, steamboat operations were impossible on the river. Union craft ascended the Suwannee on several occasions, searching for saltworks and Confederate blockade runners. After the war, there were some attempts to revive activity on the Suwannee, but the few people left were not as interested in trade, as they had other options.

Troy Spring was not destined to see peace again either, and the town of Troy's fate would not be any better than the *Madison*'s. With the ongoing forced removal of Native people, the inhabitants along the spring continued to face hardship as violence continued. Somehow, the town managed to keep things going despite the ongoing war. In 1865, the city of Troy was burned to the ground. The exact details of Troy's fate were a source of debate then and remain so today. The fire was suspected to have been intentionally set. Rumors at the time suggested the arson was politically motivated. The inhabitants of Troy included Union sympathizers and deserters from the Confederate army. Carpetbaggers were also around during this time. Freed slaves looking to settle in the area were not welcomed by everyone either. There was palpable social and racial tension, as the Confederate families felt they were better than the Union deserters and freed slaves who tried to settle in the area. While Troy would not have been the only area in Layfette County that housed Union deserters, they were part of Troy's population that did create political and social tension.

Ultimately, Troy became a town of misfits, and that did not sit well with every resident of Layfette County. The exact reasons Troy was burned to the ground may never be known. Although no lives were reportedly lost, Troy was a thriving town when it was destroyed. When the town of Troy burned, it was composed of a one-story log courthouse, five stores, two doctors, a saloon, a post office and lots of homes. Local legend suggested that the county judge was warned that the fire was going to be started and was able to remove important papers from the Troy courthouse.

Although the exact details of the fire will likely always be a mystery, the residents of Troy decided to form a new town in a better spot. They were already used to surviving harsh conditions, but rebuilding a town was still a major undertaking, even with the persistence and resilience of the townspeople of Troy. Creatively, the new town was called New Troy, and the residents began again. A building at a time, the city of Troy was given a second life.

What was New Troy like? A Mr. Sears wrote about his family history and New Troy as a descendant of one of the founders of the town. In the Sears's family history, New Troy was described as a bustling town with a gristmill, two general stores, hotels, log homes, a Baptist church (although traveling circuit preachers from different denominations would have provided the sermons) and a new two-story courthouse. New Troy was also a stop on the ferry service that made a loop up and down the Suwannee River.

Trouble in Troy, whether new or old, was not over for the Troy Springs area, however. After the end of the Civil War, New Troy went through a dark time in its brief history. Tension in the area existed between Union sympathizers, members of the KKK, freed slaves, Republicans and Democrats. A judge was assassinated in 1871, but despite everyone in New Troy knowing who killed him, the guilty person was never caught or punished. A number of murders were also attributed to the KKK during this time.

By the 1890s, New Troy had reached its peak. The town now had a cotton gin, sawmill, jail, boardinghouses and more churches. Part of the reason New Troy enjoyed growth was the steamboat traffic, with ongoing trade of turpentine, cotton, vegetables and oranges, popular items that were carried across the Suwannee. The town also supported two newspapers.

On New Year's Eve 1892, the final chapter in Troy's history began. History repeated itself: a fire broke out at the courthouse in New Troy and the buildings were lost, again. As in the first town of Troy, it wasn't clear how the fire started. When faced with the decision about what to do next, the people of Troy decided it would be best to move the county seat to Mayo, and Troy was lost—permanently, this time.

Although the courthouse was the only building completely lost in this fire, the fire was the beginning of the end for New Troy. Other buildings sustained significant damage even if they were not destroyed. One by one, the businesses left once Troy was no longer the county seat. The *Gainesville Sun* reported that houses were dismantled, and the timber, bricks and hardwood were used to build other structures in new places. Once New Troy followed Old Troy and became a ghost town, the boats that once gave life to this area soon stopped as well. Steamboat traffic to New Troy ended in 1899, and passenger ferries ended in 1917. To date, New Troy was the last human settlement at Troy Spring. Now, trees flourish along the banks of Troy Spring, the towns a distant memory.

Then, for a while, Troy Springs and Old Troy were both forgotten, and Troy Springs dropped from the public's radar. Part of what makes Troy Spring so interesting is that although it has a rich history, the spring was not easily accessible to the public once steamboat traffic ended until Florida made the purchase in 1995. Before Troy Spring became a state park, access to the spring was only by boat. This limited how many visitors could see the spring. The *Madison*'s watery grave was mostly forgotten. Although there were a few dirt roads, they were not easy to navigate and often flooded. Some local people enjoyed the spring, but it was for fishing and swimming and little used.

Today, one of the attractions of Troy Spring is that scuba divers and snorkelers alike enjoy looking at the remains of the *Madison*. Due to the cold water of the spring, there are visible remnants including metal spikes, keel rib timbers and part of the hull. Troy Spring is also not usually as crowded as some of the other springs in the area, even in summer.

Today, the footprint of the ship can still be seen in the spring run by snorkelers or scuba divers and sometimes even by those walking on the boardwalk when the water is very clear. The eerie outline of this legendary ship tells a story of prosperity and hope, violence and destruction. When at Troy Spring State Park, look for the wreck of the *Madison*, and imagine what she once was and what she would have meant for the communities along the river that waited for her whistle.

The legends of the two lost cities of Troy along with the remains of the *Madison* have made Troy Spring a hotbed of legends and spooky tales about the spring. Visitors report hearing footsteps and twigs snapping when no one is around, along with a creepy feeling of being watched. There are a few spooky options here. Perhaps it is a long-gone resident of one of the cities of Troy looking for their old town? A steamship captain come to visit his ship? Similarly to the scary stories told about Ichetucknee, one swimmer at Troy Spring claimed that he heard whispering in the bushes while swimming. He climbed out to have a look. The whispers intensified until he returned to the water.

The human story at Troy Spring is an example of how, frequently, humanity finds a marvel, and then it is lost to time and found again. Troy Spring was well known and frequented by steamships, and two towns formed on its banks. Then, when the spring was not accessible except by boat for decades, it was nearly forgotten until road improvements were made and transportation technology changed. But it wasn't until the state bought and developed the property that it became a well-known spring once again.

In North Central Florida, there is a persistent idea among locals about "hidden springs." For a long time, it was thought that there were about two hundred springs in the area. Now there is confirmation that at least seven hundred springs exist in Florida, with some estimates putting that number closer to one thousand. Some of these springs are accessible only by boat, others by hiking. There is nothing like the feeling of walking through the woods and coming upon crystal-clear water that can be enjoyed privately. Residents who know the locations of these secret or hidden springs are unlikely to share them, as these are carefully guarded secrets. Troy Spring was treated like a rumor or a spring that could only be found by the lucky few for many years. It was not exactly a secret but also not well known. It's more like the true wonder of the spring was lost to time and forgotten for a while.

With a history full of success and failure, Troy Spring remains the legacy of a time when steamboats decided the fate of the settlements they served. As a state park, Troy Spring has enjoyed its longest era of peace since before European contact, a trend that hopefully will continue to allow the *Madison* to rest in peace.

Chapter 4

FANNING SPRINGS

A Fort, a Ferris Wheel and Florida Giants

Fanning Springs State Park is located off Highway 98 in Fanning Springs, in Levy County. Fanning Springs used to be a first-magnitude spring, but due to decreased output, it is now designated a second-magnitude spring. There is an easy walking path down to the spring, and this spring is one of the few locations that is ADA accessible. Fanning Springs State Park offers swimming, snorkeling and paddleboarding.

Fanning Springs State Park has a nice, secluded swimming area. The main spring area is surrounded by concrete walls and a floating boat dock that separates the swim area from the spring run. The large swim area is deep for some smaller swimmers but has enough shallow areas for adults and older kids to be able to leisurely stand chest deep in the extremely clear, green water. The main swimming area has a nice soft sandy bottom that is easy on the feet, so swim shoes are not necessary. Swimmers and snorkelers can leisurely swim around the unobstructed swim area all the way to the deep springhead. Bring lunch to enjoy at one of the many picnic tables or venture out to the town of Fanning Springs to enjoy locally owned restaurants in the small town.

Fanning Springs State Park has both space for tent camping and rental cabins. The cabins are big enough to house a family or group. When the swimming area is open, Fanning Spring has a concessions area, picnic tables, a playground and a chairlift for anyone needing help accessing the spring water. This spring is one of the most ADA accessible.

This page: Fanning Springs. *Author's collection.*

Wildlife such as deer, hawks, woodpeckers and owls frequents the spring, and aquatic life such as turtles, bass, mullet, flounder, bowfin and manatee can also be spotted.

The spring was a source of life and survival in ages past, like other neighboring springs. Fanning Spring is part of that tradition. Archaeological sites around the spring suggest that Paleo-Indian people also lived around Fanning Spring beginning around fourteen thousand years ago. Much like the other springs, this one provided mineral water and food that were essential to everyday survival. Like Ichetucknee Spring, Fanning Spring served as a watering hole for animals. Just like at Ichetucknee, some of the animals were hunted by small groups of Paleo-Indians. Diverse groups of animals came to the spring for water, and their remains have been found in and around the spring. It wasn't just humans who hunted at the spring; animals hunted each other as well. Once Europeans encountered Fanning Spring, the evidence of animal activity and the confusion about their remains led to an interesting legend.

The arrival of Europeans in North Florida instigated a decline in the Native population as the Spanish missions used the Native people as a labor source and the mission system was used to create a buffer between the Spanish and

Fanning Spring swim area. *Author's collection.*

the English. Disease, violence and forced slave labor permanently shifted the population of the entirety of Florida, and the springs were not an exception. The Creek and other tribes in the North Florida area eventually became known as the Seminoles, and the Fanning Spring area was one of the spots where they took refuge. Fanning Spring was a Seminole stronghold because they knew the area around the spring, making them difficult to find.

For a time, the Seminole were able to keep a presence in the Fanning Spring area. They learned to raise cattle from the Spanish. With that knowledge, by the time Florida officially became a state in 1821, they were wealthy. The Seminole had adapted in the centuries-long struggle to remain on their own land and found ways to flourish. The struggle for control of the land would continue, however. With ongoing conflict between the Seminole and the Europeans, whose presence was increasing, the Seminole Wars became a way to forcibly remove the Seminole from their homes in Florida to Oklahoma. Fanning Spring was front and center in the conflict between these two groups of people.

Fort Fanning is directly across the street from Fanning Spring and a reminder of the conflict of this period in central Florida. The historic marker at Fort Fanning explains that the fort was built during the Second Seminole

Fanning Spring dock. *Author's collection.*

War. Due to the climate of Florida, the original wood buildings have been gone for a long time. The town and the spring were named after Campbell Fannin. He was charged with capturing Seminole and deporting them to locations farther west. Fort Fanning became an important spot in this effort because it was close to known Seminole villages and to the Suwannee River and a railroad crossing where captured Seminole could be transported out of the area.

Just like access to the Suwannee River was important for Captain Tucker at Troy Spring, access to the river was also key at Fanning Spring. The fort also allowed Europeans to control the Suwannee River crossing near Fanning Spring. The spring was also a dock for steamboats going up and down the Suwannee. The strategic location of the fort ensured that supplies could continue to come to the area and that transportation to and from the area could also be watched and controlled.

The troops stationed at Fort Fanning did their damage to the Seminole in the area, but their own fate was sealed. The fort was used for four to five years. Federal troops were stationed there, and during that time, thirty-one soldiers died, a large part of the total population of the fort. Their causes

Entrance to Fort Fanning. *Author's collection.*

of death varied from disease and combat wounds to drowning. All the men were buried on Fort Fanning's grounds, adding an air of solemnity to a visit to the fort. The fort was shrouded in death and disease. Survival was possible for the Seminole at Fort Fanning, but the soldiers stationed at Fort Fanning were not thriving. Eventually, the fort fell into disuse.

The hot Florida summers attracted many visitors to Fanning Springs in later years. If the springs can offer one thing all the time, it's cold water for swimmers. Picnics, camping and swimming were popular activities, but the spring area was also used for political meetings. Before there were convention centers for large meetings, there were springs. Fort Fanning was a landmark, and traveling to the spring was easier than to other locations—plus, there were entertainment options. It was a spot to meet for political and religious reasons or to have a fun day. The most well-known activity was the Fourth of July celebration, when locals and tourists alike would congregate in the spring. Watermelons were floated in the water to keep them cold. Moonshine has always been a favorite at the springs, and meals were potluck style, with people bringing food and sharing it with others at the spring. The spring became the central social hub in the area for local residents.

With steamboat traffic increasing, the popularity of the spring increased as well. For a time, there was a general store and a livery, a cotton gin and a cornmill because of the steamboat traffic. Items were sold and put on the steamboat, and other supplies were purchased to stock the stores in town. Through the years, different businesses would come and go along the spring. Some of the businesses would open just for the summer, while others tried to establish themselves permanently, such as a filling station, a sandwich shop and another restaurant. Most of the businesses were open for a short period as Fanning Springs never saw the huge boom, other than the timber industry, that other spots along the Suwannee enjoyed.

In 1905, the general store burned down, and it was not rebuilt. Eventually, the area around Fanning Spring was purchased, and a resort was built. Fanning Springs became a vacation destination complete with a Ferris wheel, a water slide and eventually a roller rink. But as much of a gem for entertainment as Fanning Spring was, even with tourists, it was never quite as popular as spots like White Spring or Ichetucknee Spring.

That doesn't mean that improvements didn't keep coming to Fanning Spring. A high dive platform once extended over the springhead. Once the state became the steward of the spring, an effort was made to make the spring area feel more natural again. The high dive was removed, and by this time, the Ferris wheel was long gone.

Four ladies swimming and eating watermelon in the Suwannee River. *Francis Johnson.*

With paranormal tourist destinations like Old Gilchrist Jail so close to Fanning Spring, it is a spot for spooky tales and worth a stop when ghost touring other locations in the area. The spring sits at a lower elevation than the parking area. The surrounding forest serves as a sound buffer, making the spring area very quiet. The silence around the spring area is eerie. There is

Fanning Springs Bathhouse. *eBay.*

also fog that sometimes rolls in from the river. The spring's nearby neighbor, Fort Fanning, is also a spot known for the supernatural, adding to the spooky vibe of the spring. Given the dark history of the fort and the number of deaths that have occurred there, it's no wonder that the location of the old fort has an eerie feel. The unnecessary deaths of the soldiers stationed at Fort Fanning combined with the fort's sinister mission and the unmarked graves left behind add to the ominous mood of this area.

The spring water is seventy-two degrees, like that of many freshwater springs; however, there is an extra-cold current that zips through the water. This may seem like a natural occurrence, but some report feeling an otherworldly cold in the spring water. In some cases, the current is so cold that swimmers never want to return to Fanning Spring again. Swimmers state that they'll feel a cold spot in the water, and then the temperature gets even colder. The reported sudden temperature drop was startling enough to one young girl who encountered this phenomenon in the water that she screamed. When her mother reached her, she also felt the extremely cold current. Some insist the water is so cold it's an entity of some kind, an ancient part of the spring. With already cold water, an even colder current could be an effective way to keep swimmers out of the spring. Whether or not the cold current is a function of the natural water system that the spring is part of or something else, it's well worth exploring Fanning Springs and

Fanning Spring entrance. *Author's collection.*

enjoying the cool currents no matter their origin, especially with a wetsuit to offset the cold.

Another legend about spots like Fanning Springs is that of the Florida giants, rumored to have had an entire civilization when Paleo-Indians would have been living at the spring. The legend is based on three things. First, very large bones have been found at Fanning Spring and other Florida springs. Like Troy Spring or Ichetucknee, Fanning Spring was a source of life and death. The clean, cold water was important to animals and humans alike. These watering holes were also a trap of sorts, as animals were exposed when they came to the spring to drink, giving both humans and other animals a better chance to find prey. For this reason, springs like Fanning Spring became the final resting place for the bones of ancient animals.

When Europeans saw some of the bones found in and around the springs, they speculated most interestingly about what the creatures were when they were alive. Keep in mind that Europeans would have been unfamiliar with much of the past and present Florida wildlife. For Europeans, the bones were a new revelation that led to speculation about what creatures' remains were left behind. Much like seeing a dinosaur skeleton and wondering what the dinosaur may have looked like, the skeletons found in Florida caused fascination and speculation. Some of the larger bones found in Florida were thought to belong to a giant race of humans believed to once live along the spring.

There was a second reason why the idea of giants roaming Florida seemed plausible to some. Early accounts of when Europeans met the Timucua all agreed on one thing: the Timucua were much taller than anyone they had ever encountered. The height of the Timucua seems to have been exaggerated in some accounts as well. There is no evidence that the Timucua were over seven feet, but they were described as being supernaturally tall. Others claimed that the Florida giants predated the Timucua. The fossils found at the springs and the legends of Florida giants led to tales of strange happening at the springs and the forests around them that were attributed to these giants.

The third reason that the myth of the giant Floridians became popular was the work of what can only be called an early "Florida man." Wanting to sell paintings and artwork, a man named Theodor de Bry claimed that he bought paintings from the 1530s depicting these giants. In all likelihood, de Bry made the paintings himself and attributed them to an earlier artist. Then, for the next four centuries, de Bry's made-up paintings supposedly depicting giant Floridians that he claimed were Timucua became the fake news that fueled this myth. If there were old paintings of these giants, they must have been real, right? Theodor de Bry certainly would not be the last Florida man to profit from making a Florida legend. His paintings stuck around as evidence of Florida giants for centuries. In this era, Florida was considered a new Eden. Some even claimed the original Garden of Eden was somewhere in Florida. That sense of magic and the ancient mystery as well as the allure of a new wilderness gave rise to the tall tales, myth and fantasy that still account for much of Florida's popularity.

Lastly, some prominent Florida characters supported this idea of Florida giants. A mayor of Tampa, D.B. McKay, stated: "A legend that is scoffed by many but given credence by others is that there were human giants in Florida in the prehistoric era." Mayor McKay supported both origin stories for the Florida giants. McKay thought the Florida giants could be from ancient times, or perhaps they were Timucua. For the Timucua version of the legend, McKay added an entire backstory, including a nine-foot-tall Timucua prince who ruled from Cedar Key. McKay called this entirely fictional person the "Last Timucuan." At that height, this nine-foot prince could easily walk to Fanning Spring or any other spot. Like anyone else, the giants needed sustenance from the springs. The truest part of this story is that, as unlikely as it was that the Last Timucuan was nine feet tall, there was, undeniably, a last Timucuan. Though they'd lived around Florida's springs for centuries, after the Europeans arrived, this civilization had about a century and a half before it would never walk the banks of a Florida spring, or anywhere else, ever again. Their height was legendary, and that actual physical difference between the Timucua and the Europeans made them a favorite of the slave industry when these so-called Florida giants were captured and put up for sale to the highest bidder.

Threats to Fanning Spring are more tangible than any of the legends surrounding it. Nitrates are causing problems to the point that, at times, the spring is closed for swimming. Swimming was once Fanning Spring's main attraction, and now swimming isn't always safe in the spring. The pollution is caused by the runoff from agriculture, livestock and septic tanks that have not been properly installed or maintained.

This spring is also an example of decreased flow, where, like Rum Island, the flow of water to the springhead has decreased. The decreased flow could be from removing water from the aquifer for bottled water, like Nestlé and other companies have done for a long time at the springs. Water is pumped out of the supply of water that feeds the spring and sold as bottled water. The problem with this system is that the water is then removed from the water cycle and won't return to the groundwater supply the way water used in other ways eventually does. The decreased water pressure through the spring not only leads to a lessened flow, but it can also cause changes to the structures that form the springs. Long term, it is not known what the outcome of bottling water will be. Short term, it has already caused issues for several springs in Florida.

Chapter 5

BOULWARE SPRING AND THE WATER WAR

Florida is a land of water: springs, lakes and the slow-flowing river also known as the Everglades. It is hard to imagine a time in Florida when water felt scarce. For most of human history, water was the key resource that dictated where a settlement, town or village was possible and if that town would make it in the long term. In a state like Florida, surrounded by and full of water, it is not hard to imagine a time when water (and access to water) shaped so much of the human story. Picking a place for a village, settlement or town has had a recurring theme: access to water. Like at Troy Spring, a spot for a town might not even be picked. Instead, access to water determined where the town went. For Gainesville, Florida, access to water became the reason it exists.

The spring was named after Dr. William Boulware, a prominent physician. Boulware Spring is a fourth-magnitude spring that sits close to Paynes Prairie and the La Chua and Hawthorne trails. The spring produces 194,000 gallons of water a day—a smaller spring, with less of a flow than others in the area. Nonetheless, Boulware Spring was an important stop for water for people heading toward the ferry crossing at the Alachua Sink. Both Union and Confederate soldiers used the spot to get fresh water. Although the springhead is difficult to see due to the murky water, it sits next to what used to be a waterworks building.

Boulware Spring was the reason Gainesville became the Alachua County seat. When the railroad was going to be extended from Cedar Key, a decision had to be made about whether the railroad would go through Newnansville

Opposite: Boulware Spring. *Author's collection.*

Right: Waterworks building. *Author's collection.*

or Gainesville. The water supply provided by Boulware Spring was the deciding factor. At a picnic, residents decided to move the seat to Gainesville based on the water provided by Boulware Spring. From 1898 to 1948, the spring served as Gainesville's main water supply.

In 1899, the Boulware Spring Waterworks building was constructed, and it still stands today. The issue of a fresh, abundant water supply became important to Gainesville's development a second time when the University of Florida decided to relocate to Gainesville after being promised free water, permanently. With the appeal of a free and abundant water supply, the University of Florida left Lake City and moved permanently to Gainesville. The University of Florida still gets free water from Gainesville.

The Boulware Spring Waterworks building has been remodeled more than once, but it still is a magnet for vandalism. Although the building is on the National Register of Historic Places, it remains in disrepair. The building is the source of several ghost stories in Gainesville. The vandalism on the sides of the building pay homage to some of the spooky tales. One

Boulware Spring vandalism.
Author's collection.

spray-painted warning on the side of the building depicts a demon-type creature and warns viewers to be gone from the area before sundown. A second spray-painted demon-like creature has the following inscription:

Deep in the merky [murky] *waters*
lays an old dream, never forgotten
but words have yet to spoil its chorus.
Awake oh dreamer, for we are your song. 2022

Now, Boulware Springs isn't much more than a hidden part of Florida's and Gainesville's history. The site is easy to miss, even with a historical road marker and access to a roadway that is often travelled by both locals going to and from work and home and out-of-town visitors going to the nearby and popular La Chua trail. This area of Gainesville is also home to the highest concentration of alligators anywhere in the world. The waterworks building is a favorite spot for Gainesville residents seeking a haunted experience, especially at Halloween, but other than that, the site is rarely visited, other than by hikers walking through to other trails.

Although the town of Gainesville would eventually outgrow the spring's ability to provide water, the flow is no longer the problem. Currently, Boulware Springs exceeds the safety limits for nutrient contents in the water. Runoff from landscaping fertilizers and septic tanks has polluted the spring to the point that its water is no longer considered safe to drink. With Florida's water supply bottled and sold to other places, the role of water in Florida's future is a story that continues.

Chapter 6

WHITE SPRING

Taking the Water and Drinking the Kool-Aid

White Spring is a dry spring. That term is a bit of a misnomer because the spring has not completely stopped flowing, and in times of higher amounts of rain, the spring's output does increase. The once swimmable spring was a second-magnitude spring with an estimated thirty-two thousand gallons a minute flowing through the springhead. The water from the aquifer flowed through limestone that contained sulfate, creating mineral water. Due to the sulfate content, the water had a rotten egg smell. Other than the odor, the water would not have been dangerous to swim or bathe in. Now, though, White Spring looks like a mausoleum of a spring.

Despite the smell, White Spring was considered a sacred spot by more than one Native American group. White Spring has gone by several different names, including Mineral Spring, Jackson Spring, Upper Mineral Spring and White Sulfur Spring. Before the spring was called by any of these names, Paleo-Indians, Timucua, Apalachee and Seminole believed the spring to have healing properties. When the Spanish wrote about the area in the 1530s, they described the Suwannee River serving as the border between the Apalachee and the Timucua. Both tribes were allowed to access to what would later be named White Spring because it was considered sacred by both. No fighting or violence was allowed by either tribe when a member went into the sacred water. The reason was that the water was thought to have healing properties, and because of its mystic and sacred nature, interference of any kind was not allowed at the spring. Even during times of war, a warrior could not be harmed while in the sacred spring. All the Native people who respected the sacred and, perhaps, healing nature of the spring were driven out of the area by Europeans, however. Once Europeans gained control of the spring area, Native people would never again live along the banks of this sacred water.

With the Native communities violently and forcibly removed to the West, the Europeans in the area started their own settlement and carried forward the belief that White Springs was sacred and had a supernatural quality. Part of the mystic reputation of the spring could have been that the mineral water was making bathers feel better. It could have improved bathers' skin, reduced inflammation (due to the temperature) and reduced stress and anxiety. At a time before modern medicine, when there was not much hope for alleviating ailments like these, White Spring offered a beautiful location in a warm climate that was easily accessible by stagecoach and, later, a train. At the end of the journey, a fun tourist town with healing waters awaited you. White Springs was the place to be and a place that could make you feel better.

White Springs' rise to a tourist destination relied on the old stories and rumors about what the Apalachee, Timucua and Seminole believed about the water having special properties. Some of the rumors were historically accurate, but others took on a life of their own. The stories were persuasive enough that in the 1830s, a wealthy plantation owner learned about the legendary properties of White Spring—and he was rich enough from using slave labor on his plantation that he was able to purchase a large parcel of land including the spring property. The plantation owners were Bryant and Elizabeth Sheffield, and they changed the course of the history of White Spring.

Bryant quickly became an advocate of the spring water, claiming that since he purchased the area and started to drink and bathe in the water, he felt better and enjoyed better overall health. Whether the spring had real health benefits or this was a marketing tactic used by Bryant, his actions did back up his claims. Bryant claimed that the spring was too therapeutic to keep to himself. He allowed public access to the spring, and a few years later, he built a simple log bathhouse around the spring. The Sheffields were the first people to market the spring as having healing properties. They advertised the spring as being able to treat nervousness, kidney disease and rheumatism. Later, the Sheffields added a hotel, and one of Florida's first medical tourism destinations was born.

Right: Historical poster at White Springs bathhouse, *Author's collection.*

Below: Drone photo of the White Springs bathhouse. *David Peaton.*

Although tourism to the spring would slow down during the Civil War and Reconstruction, Florida remained a popular tourist destination. Medical tourism picked up when "taking the waters" became popular in Florida. Tourists flocked to Florida to try out the various spring waters that had different mineral contents. Different spring waters across Florida were advertised as having different healing benefits. Journalist Craig Pittman described the appeal of White Spring:

> *Springs were Florida's first tourist attraction, drawing visitors who believed that water burbling from beneath the earth's surface must have healing properties. In the early 1900s, so many well-heeled tourists seeking cures for rheumatism, indigestion, dandruff, insomnia and a slew of other ailments flooded the little town of White Springs, north of Gainesville, that 13 hotels and a railroad line were created to cater to their desires.*

White Spring became a top destination for tourists seeking healing waters. The community around the spring was also called Rebel's Refuge for a time during and after the Civil War. Quite the opposite of the city of Troy at Troy Spring, the White Spring area was a welcoming area for Confederate sympathizers. The timber boom also drew people to White Spring. Both the timber boom in Florida and increasing tourist numbers caused the area around White Spring to grow quickly. The town of White Springs, which had a population of around one thousand residents during its heyday, saw significant growth and development. As the reputation of White Spring as having healing properties spread, more and more visitors seeking relief from various ailments showed up at the spring. Visitors could access the spring via a boardwalk that led to the natural mineral spring pool, where they could partake in the water's health benefits.

The tranquil and scenic surroundings of the Suwannee River added to the resort's appeal. At a time when the Suwannee River was a busy transportation hub, White Spring was a significant stop on that route thanks to the earlier work of Captain Tucker, who used the steamship *Madison* to prove that White Spring was reachable via steamship. Once the Suwanee River was deemed navigable by the State of Florida, traffic along the river blossomed, and White Spring was a beneficiary of Captain Tucker's work. Additionally, the town's location along a railroad line connecting White Springs to other towns along the Suwannee River made it accessible to more travelers. Eventually, railroads added an ease of transportation that made health resorts like White Spring boom like never before.

Overhead view of the White Springs bathhouse. *David Peaton.*

White Springs was popular enough that developers began to see it as an opportunity to profit. The business community saw the chance to take White Springs to the next level. The business community started to shape the town as not just a simple place to bathe and enjoy the water but a destination for more than the spring experience. The late nineteenth and early twentieth centuries were a period when health resorts were quite fashionable, and entrepreneurs banked on the springs being trendy. Individuals from all over the United States sought out destinations where they could experience the supposed curative effects of natural springs, but the business community wanted to find ways to keep those tourists spending money while they visited.

By 1903, the original springhouse had been enclosed and expanded. It was made of coquina and concrete, a popular building style in Florida. The building was designed by a famous architect in Jacksonville. The bathhouse and surrounding area were constructed to minimize water intrusion during flooding. Without this protection, anytime the Suwanee River flooded, the spring would be browned out or inaccessible until the water went back down. Sometimes it could take several weeks for the spring to be accessible again. While the spring was browned out, it meant less tourist dollars. The new bathhouse tried to address this issue by helping to control the flow of water in order to make sure the spring was always accessible. The building was three stories tall and was built around the spring. The bathhouse had doctor's offices, concessions for sale, changing rooms and an elevator. Soon, an entire spa town popped up with a focus on wellness. The popularity of

Old hotel in White Springs. *Author's collection.*

the spring and its magical, healing waters continued. More hotels were built, and other entertainment options such as a roller rink and cinemas were constructed. From a simple log cabin around the spring, White Spring became a full tourist wellness destination city with every amenity and entertainment option.

In response to the growing interest in the springs, developers built hotels and resort facilities to accommodate the influx of visitors. The eventual overdevelopment of White Springs followed a pattern repeated throughout Florida as tourism increased. The Telford Hotel was the crown jewel in the development at White Spring. Constructed in 1902, the hotel was among the most notable of these establishments. This grand hotel boasted luxurious accommodations and amenities, including a ballroom, and it became a hub for the town's well-heeled guests. The resort also became a fashionable spot for hosting social events such as dances, concerts and elaborate banquets. Popular fashion trends were sold in White Springs, but White Springs also became a force of its own. The fashionable people visiting White Springs included celebrities and other elite people. What they wore, ate or enjoyed at White Springs set trends. Everything one needed to be trendy at one of the

Front view of the White Springs bathhouse. *David Peaton.*

social events was available for sale in White Springs. Dresses, gowns, hats, boots, accessories, jewelry, wellness items and custom-tailored clothes were available for sale.

The spring and its healing waters may be what attracted people to the town, but White Springs became more than that. It was the place to go and a place to be seen. Guests could enjoy leisurely strolls along the riverbanks or attend the numerous social activities held in the hotel. The resort brought economic prosperity to the area. It was a thriving destination for travelers and a place that members of high society, the "influencers" of the time, visited. As more of these socialites visited and shared their experiences, the reputation of White Springs peaked. White Springs became so popular that U.S. presidents visited to enjoy the healing properties of the spring. For the top influencers, presidents and other wealthy or powerful people, the Telford was where they would stay.

Other hotels were built to cash in on the success of the bathhouse, but they catered to other guests. The town of White Springs ended up having fourteen hotels, boardinghouses and other luxuries. White Spring also boasted several different options for tourists on any budget. Wealthy travelers had luxury hotel options. There were also cheaper hotels and boardinghouses. White Spring was segregated, with Black hotels and boardinghouses the only option for African American travelers.

In 1906, more concrete was added around the spring to try to keep floodwaters out. The improved structure had mechanical gates that would help keep the Suwannee's surging waters at bay when water levels were

Top: White Springs bathhouse. *David Peaton.*

Bottom: Side view of the White Springs bathhouse. *David Peaton.*

high. Although it was expensive to essentially erect a giant concrete wall and building around the spring, it also made the spring a more attractive destination year-round. With the river's water level controlled by the gates, the spring water was protected, making visiting more attractive to guests, who otherwise would have had to risk the spring being inaccessible depending on the changing river water levels.

The town's and the spring's fate remained intertwined. By 1909, there were four rail stops in White Spring per day. White Springs attracted record numbers of people as the spring and town continued to gain popularity. The popularity of the spring and the diversions of the town seemed to have no limit.

Then, in 1911, a fire started in the home of Reverend Wambolt, a Baptist minister. The fire quickly spread to the rest of town. The business district was hit the hardest by the fire, which destroyed a livery, a millinery and three hotels. Thirty other structures were damaged in the fire. The fire happened at a time when many of the men in White Springs were out on a hunting trip. It was the African American community, excluded from the hunt and many of White Spring's luxuries, that saved the town's economic prospects. Working together, they stopped the fire from ravaging every structure in town. As happened with both iterations of the town of Troy, fire had the power to quickly and permanently change the future of a town. Many of the luxury hotels at White Springs were destroyed by fire, but among the few survivors was the Telford Hotel, which remained important to this spring's story.

As medical advancements and changing trends in health care took hold in the twentieth century, the popularity of health resorts like White Springs faded. People increasingly turned to modern medicine rather than natural spring waters for their health needs. The fire was also a turning point, as some businesses and hotels never reopened. Tourists did visit the spring after the fire, but never again as many as before the fire. An economic downturn hit the town of White Springs, and it never really recovered.

The widespread fame of White Springs is hard to imagine now, as swimming in the spring has been over for decades. Like the spring, the town of White Springs sits empty compared to what it once was. The spring flow slowed in the 1980s and eventually stopped. The remnants of the bathhouse can still be seen as the now dark water still swirls between the walls of the old bathhouse. This spring is also a casualty of changes in the water levels of the Floridan aquifer that feeds water to all the state's springs. Although some of the ebbs and flows at the springs can be attributed to natural causes, such as changes in rainfall, human-centered activities have also changed the health and quality of our spring water. What was once a thriving spring is now a murky, unappealing body of water that no one swims in, much less visits to expect any type of healing.

Today, some of the homes and hotels from this era survive, but not as tourist destinations. One building in particular carries White Spring's legacy. The Telford Hotel's story continued, but in a much different way than when

Suwannee River side of the White Springs bathhouse. *David Peaton.*

it was the peak of luxury. Many resorts closed their doors permanently, and even the Telford closed for a while. Over time, White Springs' resort era became a memory, and the focus shifted to preserving the town's historical and cultural heritage. Born during the heyday of White Spring, the Telford continued to evolve as the decades passed. The property has been a nursing home, an apartment building, a private home and the global headquarters of a very interesting broadcasting company.

Today, while the grand hotels that once graced the town have faded into history, White Springs remains a place of historical significance and natural beauty. Visitors can still explore the picturesque surroundings and the beauty of the Suwannee River, connecting with the past while enjoying the area's natural charm. The area has a certain spooky beauty to it. What once must have been a beautiful stroll along the river, looking at the river and the spring, now feels like paying homage to a now-dead spring. The spring is a memory of what once was, as are the hotels, fashionable events, balls, feasts and, most importantly, the healing spring water. All of it is gone, other than the remains of the bathhouse, a tomb to an era of Florida's history that remains a distant memory.

The vibe of magic, however, was not quite over for White Spring and the town around it. In the 1980s, White Spring, once seen as a sacred, magical place of healing, became a place that welcomed those seeking paranormal experiences. Some of the health claims about White Spring had always bordered on pseudoscience (although there were some legitimate health

Telford Hotel. *Author's collection.*

benefits to the spring water). But embracing the paranormal and known pseudoscience was a new development.

An example of this change was the Telford Hotel. A character named Chuck Harder and his wife purchased the Telford Hotel during this era in White Spring. There, they launched a studio for their Sun Radio Network, also known as Talk Star Radio. Later, the hotel served as a production location for a self-produced show called *For the People*. This radio and TV network was part paranormal, part pseudoscience and part sensationalism. Harder's show predated YouTube and digital cameras, so it would have been with a great deal of effort that Chuck Harder wrote, produced and marketed it. He was passionate about spreading what he considered truth.

To illustrate how out of the ordinary the content of this production was, one of the popular guests on the show claimed that there were remnants of a civilization on Mars. Chuck Harder presented the claims of guests on his show, such as the Mars author, as factual. Harder supported the claim that the lost city on Mars had a pyramid, a structure with a face on it and other ruins. For Harder, all of this was as real as the hotel that housed the broadcast. Some of the old broadcasts have now been made available on YouTube, but much of what remains are alien conspiracy theories. In some ways, White Springs and the Telford Hotel were an offshoot of some of the more outlandish claims about the magical properties of the White Spring waters.

Harder was not the last person to recognize that there was something in the air at White Springs besides the sulfur smell. Even today, paranormal and abandoned structure influencers visit White Springs and the bathhouse. The hotels and the remnants of what was once the bustling city of White Springs are also the subject of social media content around paranormal activity, including ghost stories and alien encounters. A Viking temple currently sits downtown as well. One paranormal activity influencer on YouTube even claims that the Suwannee River is a dark entity on its own and that its presence is physically discernable at White Spring—not the first time the Suwannee River has had a spooky tale told about its banks. Jenn and Ed Coleman, frequent campers along the Suwannee and travel bloggers, describe one reason why they think many spots like White Springs have failed to thrive in the long term. They recount a story frequently told at Suwannee River campfires about a curse placed on the Suwannee River valley after the final Timucua Rebellion.

Rewinding back to Ichetucknee Springs and the kidnapping of the chief's daughter: the remaining Timucua saw those events as a warning of what was to come from the Europeans. After the last stand of the Timucua during the rebellion, a chief was captured. The Colemans translate the chief's final words this way:

> *I have done as a brave man, and struggled and fought like a man until I took refuge in this pond. It was not to escape death, or to avoid dying but to encourage those who were there and had not surrendered. I ask that my people not have anything to do with these Christians, who are devils and will prove mightier than they. If I have to die, it will be as a brave man.*

According to the Colemans, these final words were the beginning of a curse that persists and haunts the Suwannee Valley. Called the Napituca Curse, it prevents any long-term successful development of the Suwannee River valley. Looking at White Springs and the empty bathhouse that once used to be a hot destination, it is easy to see how the area feels cursed, whether that origin is the Napituca Curse or not.

There isn't much to do at White Springs now except take a self-guided tour of the old bathhouse along the river. Unfortunately, as of January

2024, access to the bathhouse is restricted due to damage the area sustained during Hurricane Idalia in 2023. When visiting this historical area, however, take some time to check out Stephen Foster Park right next to White Springs. A small park dedicated to local stories, it is a great way to soak up some Florida history.

In recent times, after high rainfall, the spring has flowed again. Given time to recover from the human impact on the spring, White Springs could flow again the way it did when it was considered sacred water. Draining and bottling this sacred water has impacted White Springs. Just like Troy Spring serves as an underwater graveyard for the *Madison*, the bathhouse at White Springs is a mausoleum and the final resting place of a now mostly expired spring, a testament to how a powerful spring and a once trendy tourist destination quickly became a tomb to what was once was and is no longer remembered.

Chapter 7

RUM ISLAND SPRING

Moonshine and a Mayor

Rum Island Spring is a top pick as a starting point for a day of spring-hopping and exploring, even if it is a smaller spring. The history that Rum Island Spring has witnessed is anything but small and is reminiscent of the rumrunners that used to frequent Florida waters. First, let's get to know Rum Island Spring.

Rum Island Spring is a county-owned park inside Columbia County. Entry is five dollars per carload. Once inside the park, take advantage of two separate boat launch areas, one for larger boats and one for kayaks and canoes. There is one set of bathrooms in the park but no other facilities. Bring a picnic basket and set up some lawn chairs or use one of the picnic tables provided.

This spring is a landmark for many boaters, swimmers and kayakers, serving as a good spot to begin a day on the water. Rum Island Spring is unique in that the spring sits right next to the Santa Fe River. From the swimming area, you can touch the river and swim between both as the barricade is easy to navigate. With the river being so close, there is a tannin shading effect that happens to the color of the green-blue spring water as it flows and mixes with the brown tannin-rich water of the Santa Fe. The proximity of the spring and river water at Rum Island Spring creates a blend of the blue, green and brown waters.

Although the swimming area at Rum Island is small compared to others nearby, part of what makes this spring unique is that the swimming area is shallow, making it a good spot to take kids or less experienced swimmers.

Rum Island Spring swimming area. *Laura Barone.*

The water does get deeper at the springhead, but with the majority of the swimming area being only a few feet deep, this spring is a great option for children, although the exact depth of the spring will change based on the river's water level.

Boaters often make Rum Island a stop on their way down the river and swim in the spring before moving on to their next spring. On a quiet day, being so close to the river makes this a perfect spot to watch wildlife, and this area has plenty of birds, fish and turtles. This spring has a variety of fish, and you can see where the spring water meets the river, so it is worth bringing a snorkel.

Stopping at Rum Island for a swim while kayaking the rest of the Santa Fe River is an easy option as there is ample parking, a boat on-ramp and bathrooms at Rum Island, making it a good place to start. Rather than just a starting point, however, Rum Island is worth spending time at all on its own, beyond just being an easy kayak stop.

Being so close to the river can also be a disadvantage, as one challenge for Rum Island Spring is frequent brownouts. Typically, the flow of the spring is strong enough to push back the river water and keep the spring water clear. When the river's water level rises quickly, such as during a storm, the river water floods into the spring and makes the water brown. At times, there is a physical, floating barrier that divides Rum Island from the river. This barrier stops some of the brownouts, depending on the river's water level, but more recently, boaters have inadvertently torn off pieces of the barrier, making it a pollution issue on its own.

From Rum Island Spring, the spring's namesake is visible: Rum Island. Rum Island is separate from the spring and sits in the middle of the Santa Fe River, splitting the river in half for a short distance. From Rum Island Park, you can watch as kayakers navigate either to the left or the right of the island. Rum Island Spring has an exit/entrance ramp for boaters, making it a popular exit point for paddleboarders and kayakers. Rum Island is forested, making it a home to birds such as barred owls, ibises and osprey. Maybe it's the way the island splits the river in two, but looking at the island, there is the sense that what is immediately seen from shore is not all this island has to offer. There is more to Rum Island's history, even if there is no brown sign to tell that story.

The way the island forks the river creates chaos on the water during busier days among less experienced paddlers. That chaos mixed with the serenity of the natural setting of the island is a metaphor for the history of this beautiful yet unruly space. Although Rum Island Spring and the park

This page: Drone photos of Rum Island Spring and the Santa Fe River. *David Peaton.*

that surrounds it are now a family-friendly, alcohol-prohibited space, that hasn't always been the case. Rum Island has a notorious history that begins with Prohibition.

Florida has a rich history of rum-running during Prohibition and after. There are legends about limousines lining up along the wharf in Miami Beach to meet boats carrying rum. Rum Island is part of this legacy of illegal bootleggers. The movement of rum and other liquors up and down the Santa Fe River was a known activity as far back as when Native Americans populated the area and furs and other Florida goods were traded with Cuba for rum and other liquors.

Building on that legacy, Rum Island presented a perfect opportunity for bootleggers. Rum Island Spring provided an endless, easy supply of fresh water. Additionally, Rum Island offered something that no other spot could offer bootleggers. Although Rum Island is under state jurisdiction for law enforcement since it is on a waterway, the island is also positioned outside the jurisdiction of neighboring Alachua, Columbia and Gilchrist Counties. Because its remote location made enforcement difficult and expensive, it was not especially worth it for law enforcement to police the area. It became a spot outside the law, offering an ideal location for illegal stills. The location of Rum Island also worked well because the rum could be easily moved down the river to other locations in and out of Florida.

Certainly, there were more famous bootleggers who frequented Florida waters, but on a local level, Rum Island was notorious for smaller stills that would be transported to other areas of Florida via the river. Some of the bootleggers had quite an impact on surrounding areas such as High Springs. One local bootlegger who had a lasting impact on High Springs was Juanita "Skeet" Easterlin. Building on the tradition of local Rum Island bootleggers, Skeet Easterlin took local moonshine to the next level. Wearing overalls at a time when female students at the nearby University of Florida were not permitted to even wear pants, Skeet stood out immediately. Always holding a lit cigarette at the end of her long cigarette holder, Skeet was a hotshot who did her own thing, her own way.

The moonshine operation was not Skeet's only "secret" that was known by everyone in town. She shared her home with thirty-five cats and traveled to Jacksonville frequently to see a woman known to be her lover. Skeet lived life on her own terms. Her choice of clothes was an immediate visual indicator to anyone who saw her that she was different and didn't care about social expectations. Her clothes were practical for her everyday work, and for Skeet, that was what mattered. A woman owning a business, wearing pants and supporting herself financially would have all been tea for the town gossips, but people whispered about Skeet's love life, too. Skeet had a long-term lover who lived in Jacksonville. Marriage would not have been an option for the two women during this period, but that barrier did not impact Skeet's love and lifelong devotion to her partner. The town knew about Skeet's relationship, and whether or not they approved, the citizens of High Springs still bought Skeet's gasoline and her moonshine.

Left: Spring sign. *Author's collection.*

Below: Swimmers at Rum Island Spring. *Author's collection.*

Legends about Skeet are about as plentiful as springs in Florida, but there is no doubt that Skeet and her business ventures had a lasting impact on the area. Although Skeet would eventually become High Springs' first female mayor, she also made her way as an entrepreneur at a time when women were few in the business world. Sitting in the heart of downtown High Springs where the present-day Great Outdoors Restaurant stands, Skeet's Corner was a center point for High Springs history.

Skeet's Corner was an everyday stop for locals and visitors to High Springs. Skeet's Corner was primarily a gas station but also a café, the High Springs bus stop and, at one point, an appliance store. The spot was known for home-cooked food and exceptional coffee. Skeet Easterlin had a plan for Skeet's Corner, and part of her success was her marketing, as ads for Skeet's Corner were active in local newspapers and magazines. She was aggressive, and it worked: Skeet's Corner's reputation spread across North Central Florida.

Skeet's Corner became successful, and Skeet was able to support herself. But that was not quite enough for Skeet, and her corner held a widely known secret. If a patron pulled into the front of the gas station, the legal options were available (gas, food, merchandise). If a customer was feeling more adventurous and pulled around to the back of the gas station, Skeet's moonshine was available for purchase.

Skeet's Corner. *High Springs Historical Society.*

Skeet continued to develop Skeet's Corner as the town's needs changed, but it was her moonshine operation that took off. Late at night, Skeet made her moonshine using what was a small still at first. Remote locations like Rum Island were a smart choice: Skeet knew that law enforcement might not be as motivated to shut her operation down because these remote locations were difficult to search. Partly due to financial issues, as her expected inheritance was tied up in court for a long time, moonshine became a way for Skeet to continue to support herself.

While making moonshine available at such a public, popular spot in High Springs, Skeet also needed to protect her operation. Cleverly, she purchased all her supplies from the other business owners in downtown High Springs. She likely could have gotten a better price on ingredients like sugar for her still, but by buying from local shop owners, she ensured their silence. As she was a steady customer helping to keep their businesses afloat, they had no motivation to turn Skeet in. The other businesses needed Skeet to keep coming back and her customer base to continue to grow. Her success was their success.

Despite Skeet's reputation as a ferocious woman, the townspeople of High Springs respected her. They recognized that she was direct and ruthless at times, but they also knew she was fair and could be generous. Being a businesswoman by day and a moonshiner at night was not quite enough of a challenge for this eccentric woman, so when the townspeople pushed her to become mayor, she set her eyes on holding public office. Ironically, her campaign centered on stricter enforcement of illegal alcoholic beverage sales. Skeet was going to win mayoral office by offering a solution to the very problem she contributed to in the High Springs area—a political philosophy still used in Florida today.

Public office was a turning point for Skeet. She learned quickly that Florida voters can be fickle. Although High Springs locals asked her to run to "shake things up," it turns out that they did not actually want change. The citizens who urged her to run almost immediately turned to opposition when she tried to move the police chief to night duty, a demotion she felt was warranted. Skeet justified this move by claiming that the police chief was not effective. Certainly, in Skeet's case, he wasn't. She had been making and distributing moonshine for years, and he had not been successful in shutting her down. Ultimately, the commission overruled her decision, and in a later letter to the community, Skeet said, "If you think the police department is as effective as it could be, you should have told me." Skeet gave up on removing the police chief, but she made a lifelong enemy in the process.

Skeet's Corner stood at a time when High Springs was still fully segregated, when restaurants only allowed Black patrons to come in the back door and water fountains were labeled "White only." The famous and long-standing Priest Theater, for example, had a balcony reserved for Black patrons, as they were not allowed on the ground floor.

High Springs was completely segregated, meaning separate entrances and seating areas for Black people in restaurants, bathrooms and so on. It was a dangerous time for Black Americans in High Springs, so much so that Black schoolchildren were instructed by their parents not to walk home through downtown whenever possible. Instead, Black students walked the long way through the woods, down a remote road, to avoid the downtown area and the trouble interacting with White residents could bring. Black children were told, if they did have to walk through town, not to stop or bother anyone.

Longtime High Springs resident Roger King recalls breaking that rule once during the height of segregation in High Springs. He remembers a day when he and a friend were playing around on the way home from school and got thirsty in the Florida heat. Roger made the risky decision to go to Skeet's Corner. He still remembers his parents' guidance: "You know, they wanted to keep us safe, so when you're going to school, you're just going to school. Walk on through town and don't bother anything and come on back home." On this day, however, Roger and his friend decided to take the sidewalk directly through downtown High Springs.

Thirsty from playing around, Roger approached Skeet's Corner. Not seeing Skeet around, Roger decided to go into the gas station to get some water. He ignored the "White only" label on the water fountain and got some water, climbing carefully onto the overturned soda crate used as a stepping stool to reach the water fountain. Just as he was about to take a sip, Skeet came up from behind and lifted him off the crate. Roger started to yell at Skeet to put him down. After flinging him around some, Skeet did set the child down, but not before saying, "Boy, you know you don't drink from this fountain! You know it's White only! Now, if you want some water, you go over there and you get you one of them bottles there. And we'll run you some water in the bottle."

The bottles offered for Black patrons were not clean bottles, so Roger replied, "I don't want your old stank water!" He ran away, but the incident wasn't over just yet. Everyone knew everybody in High Springs. Skeet walked over to the dry cleaner downtown where Roger's uncle worked and told his uncle what happened. Roger was warned that doing things like drinking from a White fountain causes problems and should be avoided.

This was not the last time Skeet's Corner was the backdrop to serious racial tensions in High Springs. A White woman in High Springs claimed a man had whistled or winked at her. Everyone in town knew that this was not true. That didn't matter. The woman's husband decided he was going to take action despite the townspeople's opinion that the incident never even happened. The husband shot the accused man—the innocent man—at the main stoplight in High Springs in the middle of the day.

Not only was an innocent man murdered, but the husband also had the body laid out in front of Skeet's Corner all day. Roger King remembers, "They let him lay up out there for about all day, as a lesson to the other Black people." In a disturbing but not surprising move, the woman's husband later became a police officer. No one was ever prosecuted for the murder, and High Springs moved on like nothing ever happened. Skeet's assessment of the police force had been correct.

For Skeet Easterlin's attempted second term, her campaign once again focused on a platform of getting even stricter with operators of illegal stills and coming down harder on those producing illegal alcohol. This time, the citizens of High Springs were skeptical. As mayor during this period, Skeet also served as a judge—and as a judge, Skeet had acquitted someone caught making moonshine. The majority of the seven hundred voters who cast a ballot in 1958 decided that Skeet was no longer the town's best option for mayor.

Her legitimate reign over High Springs having ended, Skeet decided not to let formality and a little thing like an election stop her plans for the city. After losing the title of mayor, Skeet decided to continue to run the town—only this time, no longer giving any credence to legality. Almost overnight, she became known as the "Cracker Queen" of High Springs, a nickname residents quickly adopted. Skeet decided to focus on expanding her bootlegging empire.

Skeet had production up in no time, producing hundreds if not thousands of gallons of hooch per day. She moved operations to a larger location to produce more moonshine. She'd outgrown spots like Rum Island that could hide only small stills. She found a new location, but her next challenge was labor. Skeet needed an ever-growing workforce to support her empire. Though her vision was grandiose, Skeet was a small-framed woman who could only do so much of the manual labor herself. Even when her operation was smaller, Skeet was known for her ability to find and hire able-bodied men. She continued to build her operation with carefully selected allies—the backbone of her kingdom was the loyal folks who worked for her.

Rum Island Spring. *Author's collection.*

Despite the racial injustice that happened at her property and her enforcement of segregation at her business, Skeet frequently hired African American men to guard her still around the clock. One of her most well-known employees was Big Boy Koon, a particularly strong man known for his large, muscled body. While he was officially employed as a gas station operator, it was also known that Big Boy Koon was Skeet's ever-present bodyguard. She also employed women, especially for distribution. Women, Skeet reasoned, were less likely to be suspected of an illicit activity like selling moonshine. This is part of what inspired the loyalty Skeet enjoyed from her employees. She gave people a chance whom others overlooked. Due to her ability to scale up her workforce, the mayor turned Cracker Queen continued to grow both her payroll and her production and distribution network. As her network grew to *Breaking Bad*–like heights, Skeet worked toward making herself and her employees financially secure. At the peak of her operations, Skeet had men and women on her payroll all over the state, as far as Miami. Her workforce primarily helped Skeet distribute her moonshine, all produced in a dry county.

With much of High Springs either working for Skeet or being one of the business owners who sold her supplies, Skeet's position as a moonshine producer and distributor seemed secure. Then, in 1958, a scene unfolded at Skeet's Corner that had happened many times before. A stranger showed up at Skeet's Corner. It wasn't unusual, given Skeet's reputation and that of Skeet's Corner as a good place for travelers to stop. Since Skeet's Corner was located in the center of town, at the top of the hill, visitors tended to gravitate toward the property. A stranger to High Springs stopping at Skeet's Corner certainly was not an unusual event on its own.

As she often did, Skeet decided to give this particular newcomer a job when he asked for it. The man claimed he'd heard that Skeet's Corner was always hiring. As she had with so many others, Skeet hired the man on the spot, and Tom Anderson began working for Skeet. First, Tom worked in Skeet's legal operations at the gas station. Tom proved himself invaluable to her gang quickly, and soon he was overseeing whiskey operations. Anderson didn't just help manage the production of the whiskey, but he also made it a point to learn and understand Skeet's distribution system, which had grown into a statewide criminal organization.

Tom Anderson, however, had a secret of his own. He wasn't a simple newcomer to High Springs looking for a job. He was an undercover unlicensed beverage agent. Anderson was assigned to High Springs to learn more about Skeet's operations and how her distribution network worked. Tom Anderson had been working for Skeet with the goal of taking down her kingdom.

Based on the information Tom acquired during his undercover assignment, there was more than enough evidence to prosecute Skeet. Wanting physical evidence to make the conviction a sure thing, the High Springs police chief—whom Skeet had tried to assign to night duty when she was mayor—saw an opportunity to get even (and fulfill his duties). The police chief was ready, at last, to get his revenge. Despite Skeet's efforts to keep the business world of High Springs dependent on her, the loyalty of the other business owners in High Springs was not enough to keep her operation safe this time. To Tom Anderson's damning evidence, the police chief added his own insight: he knew the location of Skeet's still. Acting like the location of the still was new information he had just learned, the police chief led federal agents to Skeet's "secret" still.

Skeet was arrested for manufacturing, storing and distributing moonshine whiskey with a $1,000 bond. Two of her employees were also jailed. Federal agents seized 130 gallons of moonshine (about twice the volume of a bathtub). Her stills were producing significantly more than this at the time, but Skeet's employees had managed to hide and get rid of some of the moonshine,

Tree at Rum Island Spring. *Author's collection.*

once they learned the operation was busted, in hopes of a lighter sentence for Skeet and the members of her crew who were also arrested. The agents blew up Skeet's still, making sure that her employees would not continue the business without her. The police chief worked with federal agents to make sure Skeet was permanently shut down, and ultimately, he was successful.

Skeet Easterlin got eighteen months of jail time. She did not turn in anyone else who had been in the business with her. Despite every tactic the federal agents used to try to get Skeet to identify the rest of her organization—including those businesses that knowingly sold her supplies, her employees who helped with production and distribution and her customers—Skeet remained silent. When she returned to High Springs, she got a hero's welcome. The town and her former employees and customers appreciated that she had protected them so no one else had to go to jail.

Although financial problems and eventual dementia haunted the last chapters of Skeet's life, she left her home—known as Skeet's Cottage—to a neighbor on the condition that the property be restored based on Skeet's instructions. That property still exists today in downtown High Springs, currently a bed-and-breakfast called the Grady House. This property has been an important piece of real estate in High Springs for a long time. Starting as a bakery, the property eventually became a railroad boardinghouse and then, finally, the mayoral home.

In more recent times, Skeet's Cottage (Grady House) is not only a bed-and-breakfast but also a destination for those looking for a place with a history that may still haunt the building. The house that used to be home to a legend is now a legend for another reason.

According to the *Gainesville Sun*, a former owner of Grady House described a chessboard in the hallway that remained set up and ready to play. The pieces reportedly moved without anyone being nearby. When the pieces were put back in their original spots, they would inevitably get moved again. At first, the owner thought that a guest was moving the pieces but then noticed that even when the Grady House had no guests, the same pattern would repeat with the chess pieces.

Over time, the legend grew, and frightened guests would report a female ghost appearing in certain rooms. The owner felt sure that this was the ghost of Skeet Easterlin coming back to check her moonshine stills once again. A woman in an old-fashioned nightgown with the scent of lilacs or orange blossom is often seen in the Red Room. Some suspect that Skeet is also responsible for moving the chess pieces on the hallway chessboard. Skeet was a woman larger than life, so it is no surprise that well after her death, her legend haunts High Springs in a different way.

What does the future look like for Rum Island, the Grady House and the other spots Skeet Easterlin used to and (may still) haunt? Harvesting water for moonshine and making it in stills on Rum Island did not create lasting harm. The same cannot be said for the current practice of harvesting water

Above: Grady House, High Springs, Florida. *Author's collection.*

Opposite: Rum Island Spring water mixes with the Santa Fe River. *Author's collection.*

from the Floridan aquifer in the volume that bottling plants and agriculture currently consume. A silt barrier was recently removed and replaced to help protect the spring.

A spring that shaped the area around it, rich with a unique history, now has fewer and fewer days when the springhead is visible and the water is clear. Rum Island Spring could be in a last chapter as bleak as Skeet's final days. Rum Island Spring is in better shape than Boulware or White Springs, but this small spring is at risk of following their example. Preserving the legendary spring would mean taking a more aggressive approach to protecting's Florida's spring waters and challenging the removal of spring water for corporate profit. Rum Island Spring needs help to stay a character in the unfolding Florida story.

Chapter 8

SILVER SPRINGS, PARADISE PARK AND SPRING SEGREGATION

Silver Springs is a collection of thirty springs that push five hundred million gallons a day through sixty-one vents. The largest vent is called Mammoth or Mainspring and is around thirty feet deep, providing half the waterflow that eventually joins the Silver and Ocklawaha Rivers. Silver Springs is a hot spot in Florida's history; as a walk through the Silver River Museum shows, this region mattered to both the giant sloths that came to the spring for water and the tourists who came for entertainment.

Although Silver Springs no longer allows swimming in order to preserve the integrity and beauty of the area, canoeing and kayaking along the expansive spring run are still allowed. The main park offers shopping—or take one of the legendary glass-bottom boat rides that are available for an extra fee. While there, stop in the other section of the park and take a tour of the Silver River Museum to see the history of the park. The museum is open on Saturdays and Sundays, and as of January 2024, it is two dollars per person (cash only).

For at least ten thousand years, humans have been present at Silver Springs. The clear water would have been its own valuable resource, but the water also provided food. Today, the spring is still home to many species of birds and fish, otters, manatees, alligators and non-native monkeys.

Although other springs like White Spring were said to have healing properties, the Timucua considered Silver Spring a sacred spot. The clear, glistening waters were revered as a source of renewal. Silver Springs was a place where the natural and the supernatural converged.

Silver Springs. *Author's collection.*

With the arrival of Europeans, Silver Springs took on a different role. In the mid-nineteenth century, the area saw the construction of a small amusement park, and steamboat tours were introduced to showcase the breathtaking beauty of the springs. As time progressed, the park evolved, offering a range of entertainment options, including glass-bottom boat rides that allowed visitors to observe the underwater world without getting wet.

About a mile down the road from the Silver Springs entrance was the "For Colored People Only" sign designating the entrance to Paradise Park. Run by the same owners, Paradise Park had similar tourist attractions to Silver Springs', including glass-bottom boat rides, jungle cruises, a dance pavilion and a beach. At the time, there were only two other beaches that allowed African Americans. Although Paradise Park was supposed to be an equal facility to Silver Springs, the facilities were nicer at Silver Springs. Paradise Park was still a popular tourist destination, however, enjoying one hundred thousand visitors a year during the twenty-year period that it was open.

Staying true to the spiritual component of Silver Springs, churches held mass baptisms at Paradise Park. Paradise Park became a staple in the social lives of local residents as well, with church picnics held there regularly and appearances by Santa Claus in a glass-bottom boat during Christmas. There was also an annual beauty pageant for Miss Paradise Park.

Left: John and Cora Geertsema visit Silver Spring, 1984. *Author's collection.*

Below: Paradise Park. *Silver River Museum Display.*

Silver Springs also boasts a unique connection to the entertainment industry. The crystal-clear waters and lush surroundings caught the eye of filmmakers in the early twentieth century. Numerous films and television shows were shot on location at Silver Springs, utilizing the picturesque backdrop for scenes that required a touch of natural magic. The springs served as the setting for underwater sequences in movies like *Tarzan* and *Creature from the Black Lagoon*, further cementing their place in popular culture.

A legend that persists today about Silver Springs is about the monkeys, a story that is part legend and part based on lived experiences at the park. Boating through Silver Springs State Park or onto the Silver River, it is common to see one of the hundreds of rhesus monkeys that call the park home. Although this colony of monkeys has been in and around the park for decades, its origins are a local legend.

A common belief persists that the monkeys at Silver Springs were released during the filming of *Tarzan*. The story goes that to make the area look more authentic, like a jungle, the producers released monkeys into the forest. During filming, some of the monkeys escaped and were able to survive and persist in the park.

How the monkeys actually came to Silvers Springs has nothing to do with Tarzan, however. Instead, as with many unique Florida stories, a man decided to add to Florida's landscape. A longtime tourist entertainer, Colonel Tooey, was committed to making sure tourists were enthralled with his boat tours. Fascinated tourists lead to higher tips.

When Captain Tooey was a longtime boat captain on a lake in another state, he heard about a story that local people living near the lake liked to tell. According to legend, a phantom Native American would appear out on the lake. Tooey grasped onto the local legend of the phantom Native American and embellished the story for the tourists on his boat. Every time Captain Tooey told the story, he added more details to the lore. Tooey saw his chance to drive more tourism to the lake and increase his income in both boat fares and tips. With an accomplice, Tooey orchestrated the "Native American's" appearance. He'd have his partner dress us as a ghostly Native American and go out either on the banks of the lake or in another boat. Since Captain Tooey knew where the phantom was going to appear, his boat always got the best view of the apparition. Tooey continued with his deception because it worked. His boat tour was the most popular, and he made the most money.

Once he moved to Florida, Tooey became a captain leading the popular boat tours at Silver Springs, and he wanted a way to make his tours on Silver River more interesting. Leaning on his deception from when he was a lake

Descendant of the original monkeys released at Silver Spring. *Laura Barone.*

captain, Tooey decided to embellish the jungle theme at Silver Springs. He bought six monkeys and released them on what he named Monkey Island, convinced that the monkeys would stay at their island home. What Tooey either didn't know or didn't consider was that monkeys can swim. Although

112

not native to Florida, the monkeys flourished in the Florida climate and became permanent residents at several locations in the park. Tooey's deception worked, and visitors did like to see the monkeys and tipped Tooey accordingly. Like Theodor de Bry with his deception about the art depicting Florida giants, Captain Tooey became the next Florida man to sell a fantasy. He added to an illusion that people loved to believe: that Florida was a wild, untamed, unexpected land where anything could happen. Whether that meant monkeys on a jungle cruise or bones thought to be from giants, Florida was enmeshed in fantasy once again.

While Silver Springs continued to attract visitors and filmmakers, it also faced challenges associated with increased tourism and development. As more people visited the springs, the delicate ecosystem began to show signs of stress. Water quality issues and invasive species threatened the once pristine environment, underscoring the need for conservation efforts.

Recognizing the importance of preserving this natural wonder, various organizations and local authorities embarked on conservation initiatives to protect and restore the springs' health. These efforts included measures to improve water quality, manage human impact and safeguard the habitats of the diverse species that call Silver Springs home.

In 2013, Silver Springs took a significant step forward in its conservation journey. The State of Florida acquired the land and the attractions associated with Silver Springs, transitioning its management to state ownership. This move was aimed at enhancing conservation efforts and ensuring that the natural beauty and historical significance of the area would be preserved for future generations to enjoy. After all, the monkeys need a place to swim.

Chapter 9

WEEKI WACHEE AND
THE FLORIDA MERMAID

There is a certain type of mermaid that is native to Florida waters. Florida mermaid stories are wrapped in a legend that sailors saw mermaids in Florida waters. Weeki Wachee took this legend and created an entire industry that permanently influenced Florida culture and the perception that the rest of the country has of Florida as a land of wild adventures. While the mermaid shows are undoubtedly the star attraction, Weeki Wachee Springs offers a wide range of outdoor activities that allow visitors to connect with nature and explore the breathtaking surroundings. Some of the most popular activities include glass-bottom boat rides, swimming, canoeing, kayaking, a water park and the mermaid shows.

The Weeki Wachee River is a paddler's paradise. Canoe and kayak rentals are available, allowing visitors to glide along the crystal-clear waters, enjoying the tranquil scenery and observing wildlife. Paddling down the river is a peaceful and immersive way to experience the natural beauty of the area. Weeki Wachee Springs State Park features a designated swimming area where visitors can take a refreshing dip in the cool, clear waters of the springs. The constant temperature of the water makes it a pleasant escape from Florida's heat.

This spring is a great place for families. Other than the amazing natural spring, during the spring and summer seasons, Buccaneer Bay is an added bonus. A small water park located in Weeki Wachee Springs Park is a fun way to experience the park. Buccaneer Bay offers two large water slides, a concession stand and a seasonally open restaurant. Although Weeki Wachee

Aerial view of Buccaneer Bay. *State Archives of Florida, Florida Memory.*

Mermaids in a fish tank, 1948. *State Archives of Florida, Florida Memory.*

Spring is open year-round for swimming, Buccaneer Bay does close during the colder months, so plan accordingly. Outside of swimming, guests can either bring their own canoe or kayak or rent one from the onsite rental service and enjoy boating the spring run. Reservations are recommended, as there is a limit to the number of boat launches allowed each day. Small marinas can be found in the areas around Weeki Wachee Springs that also rent canoes, kayaks and even small motorboats.

Birdwatchers and nature enthusiasts will find ample opportunities to see the diverse plants and animals in the area. The riverbanks and forests surrounding the springs are teeming with wildlife, making this an ideal spot for wildlife photography. Several well-marked nature trails wind through the park.

Although Ichetucknee Spring was the first Florida spring to become internationally known, Weeki Wachee was one of the first to become a national sensation. Weeki Wachee Springs is nestled on Florida's Nature Coast, approximately forty-five miles north of Tampa. The primary draw of the area is the Weeki Wachee River, which originates from the springs and flows westward into the Gulf of Mexico. The river is pristine and serene, winding through lush subtropical forests and providing ample opportunities for outdoor recreation.

Weeki Wachee Springs is one of the deepest freshwater springs in the United States, with depths plunging to over four hundred feet underground. The water is incredibly clear and boasts a constant temperature of 74.2 degrees Fahrenheit (approximately 23.4 degrees Celsius), making it an ideal habitat for a variety of aquatic life.

Cypress and oak trees dominate the surrounding landscape, providing shade and creating a picturesque backdrop. Birdwatchers will be thrilled to spot an array of avian species, including ospreys, herons and ibises. Manatees, Florida's beloved marine mammals, are often seen in the river, particularly during the winter months when they seek the warm waters of the springs.

The natural wonder of the spring was enough to draw tourists; however, Weeki Wachee had its own version of Captain Tooey at Silver Spring. Newton Perry had a vision for Weeki Wachee. Inspired by the existing legends of mermaids in Florida's waters, he thought, *Why not make one of these legends a reality?* As part of this vision, Perry had a submerged theater built, as any Florida man would. Perry wanted the submerged theater to include live underwater shows, unlike any other "stage" at the time. His vison was to take the idea of Florida as a spot to experience fantasy to new levels of perceived realism.

To achieve this vision, Perry and his team developed a system of air hoses and hidden air pockets that allowed the "mermaids" to breathe underwater. Dressed in colorful costumes, these talented swimmers would perform synchronized routines and interact with marine life, creating a mesmerizing and enchanting spectacle for visitors. The mermaids would quickly become the main attraction, captivating audiences and gaining recognition worldwide. They would steal the heart of a nation that, in the years following World War II, was ready for an escape from reality.

Once his underwater stage was finished, Perry offered mermaid shows with live performances by actors and actresses using the one-of-a-kind theater. Guests would descend a staircase to the underwater viewing area. The shows were viewed via a large glass window. Decorations hid the air lines and vents that the mermaids used to make the show more realistic.

To bring his vision to life, Perry needed skilled swimmers who could perform underwater. He held auditions and eventually selected a group of women to become the mermaids of Weeki Wachee. These women were, of course, not actual mythical mermaids but highly trained swimmers capable of performing synchronized routines, ballet and other mesmerizing acts beneath the crystal-clear waters of the springs. Florida's version of the mermaid was born.

The mermaids dressed in glamourous costumes and were equipped with hidden air hoses and air pockets that allowed them to breathe underwater while performing. The costumes were Hollywood quality and continued to evolve as the show did. The appeal was not just in the elaborate costumes, however, but also in how the mermaids immersed themselves in their characters. Their ability to move gracefully underwater and interact with marine life, such as fish and turtles, created a spectacle that truly captured visitors' imaginations. Journalist Craig Pittman described the mermaid show this way, "The water was so clear that watching the show was like peering into a strange dreamscape where weightless beauties performed a languorous ballet." The show was a technological marvel at the time, and that coupled with the successful cosplay of the mermaid actresses made Weeki Wachee a success.

The shows typically featured synchronized swimming routines, underwater ballet and other stunning performances that highlighted the natural beauty of the springs and the enchanting abilities of these Florida mermaids. Over the years, the shows evolved to incorporate music, storytelling and advanced underwater technology, enhancing the overall experience for the audience. Florida became the only place in the world where "mermaid" was an actual job description. Perry knew how to market his mermaids as well, continuing to develop their increasing popularity. Pittman described the ongoing success of the mermaids: "To promote the attraction, Perry snapped photos of his cast in all sorts of unlikely underwater poses: having a picnic, hanging out with an enormous seahorse, hugging Santa. Then he sent the photos to newspapers around the country, which gladly gave him free publicity." Distributing the photos nationally did its magic. The Florida mermaids were a sensation.

The success of the show made the spring a revered tourist attraction and one that helped develop Florida's reputation as a land of fantasy, a place where wonder and magic still existed and could be experienced. Pittman described the success of the mermaids at the spring: "Tourists got hooked on Perry's aquatic erotica. By the 1960s, Weeki Wachee was attracting half a million visitors a year to watch the carefully choreographed mermaid shows." Perry picked women who were gorgeous, and he knew how to maximize the sex appeal of the mermaids and capitalize on this idea of fantasy becoming

Miss Underwater contest, 1946. State Archives of Florida, Florida Memory.

reality. Those tourist numbers were impressive and exceeded anything Perry could have imagined when he set out to create Florida mermaids.

While the shows are entertaining, they also serve an educational purpose, and the park has certainly worked on shifting toward a more educational vibe. The mermaids often incorporate messages about conservation, ecology and the importance of preserving Florida's natural wonders into their routines. This unique combination of entertainment and education sets Weeki Wachee Springs apart as a place where fun and learning go hand in hand.

Visitors to Weeki Wachee Springs can also meet and greet the Florida mermaids after the shows, providing a memorable and interactive experience. These encounters allow guests to ask questions, take photographs and even receive autographs from the talented performers.

In 1959, Weeki Wachee Springs was purchased by the American Broadcasting Company (ABC), which further expanded its appeal by featuring it in television specials and making it a prominent part of Florida's tourist landscape. The park continued to evolve, adding more attractions and amenities to enhance the visitor experience while preserving the natural beauty of the springs. Over the years, Weeki Wachee Springs faced its share of challenges, including changing ownership and financial difficulties. Despite these hurdles, the mermaid shows endured, captivating generations of visitors. In 2008, the State of Florida took over the park.

The story of Florida mermaids does not end with Newton Perry. With recent interest growing in cosplay as a hobby and sometimes part of an entire lifestyle, Weeki Wachee is no longer the only place to spot a Florida mermaid. Using elaborate costumes, tails included, and practicing advanced breath-holding techniques, mermaids are often spotted in other springs across Florida. The popularity of mermaid cosplay has led to an entire TV show dedicated to the people who make Florida mermaids a frequent sight in Florida waters.

The ecology of Weeki Wachee Springs is a testament to the delicate balance between human interaction and environmental preservation. The springs' crystal-clear water is not only visually stunning but also ecologically significant. The aquifer that feeds the springs is part of a vast network of underground water sources, making it essential to protect the water quality of Weeki Wachee Springs for the health of the entire region.

Weeki Wachee Springs State Park has been committed to conserving and preserving this natural wonder. Its efforts include monitoring water quality, minimizing pollution and creating designated swim zones to reduce the impact of human activities on the springs. The park's ecological initiatives extend to educating visitors about the importance of preserving Florida's natural resources.

The Weeki Wachee River, which flows from the springs, is an important habitat for numerous species. It is home to a diverse range of fish, reptiles and amphibians. Kayakers and canoers often paddle along the river, taking in the lush environment and spotting wildlife along the way. The riverbanks, covered in thick vegetation and shaded by towering trees, provide a safe haven for countless creatures.

Manatees are particularly iconic residents of Weeki Wachee Springs. These gentle marine mammals often enter the river in the winter months, seeking refuge from the colder waters of the Gulf of Mexico. The manatees at the park are a good reminder of where the mermaid legend started. The park and its rangers play a vital role in manatee conservation, providing protection and raising awareness about the importance of preserving this unique ecosystem. As far as preserving the Florida mermaid, that is expressed best in the mermaid's own theme song, "We've Got the World by the Tail":

We're not like other women
We don't have to clean an oven
And we never will grow old
We've got the world by the tail!

Chapter 10

LILY SPRING AND FLORIDA'S LAST HERMIT

Lily Spring is a private, oval-shaped spring with six vents that fill the spring pool and a swimming area that averages about six feet deep. Lily Spring is located near Poe Spring off the Santa Fe River near High Springs. The spring is remote and can only be accessed either by paddling up the spring run or by hiking through the woods. Lily Spring has an important connection to an old Florida legacy.

Although this is one of the less talked-about pieces of Florida history, several corners of Florida serve as home to a solitary eccentric, or hermit. Florida has a rich tradition of hermits who have shaped the uniqueness of the state and are a different facet of the "Florida man" legacy. The website Hermitary defines hermits as "persons who live alone, away from other people or have abandoned society—perhaps for religious reasons." That doesn't necessarily mean that hermits dislike people or are incapable of living near them. Instead, being a hermit is their chosen lifestyle. The landscape and climate of Florida promote the possibility of hermits wanting to live a life of seclusion or, in some cases, a self-sufficient lifestyle.

There are several examples of Florida hermits: one such is Arthur Leslie Darwin, who lived at Possum Key. Arthur Darwin selected a life of seclusion, and even when the key became a national park, he continued to live there. His everyday life was simple, with no electricity or running water. He collected rainwater for drinking, his food was all grilled and he grew fruit and raised rabbits. Twice a month, Darwin would bring his extra produce to a local produce stand and trade it for supplies he needed. Although his way of living may seem rough to the average person, Arthur reportedly lived to be 112.

Arthur Darwin made his choice to live this lifestyle simply because it was his own way. He wasn't interested in knowing what was happening in the world, even though he had a radio—which went unused. He didn't offer advice to visitors or profess to possess any deep wisdom about how to live life; he simply liked living on the key. At the end of his life, he expressed the wish to leave the key since everyone he once knew was gone. However, he lived on the key for the remainder of his life.

At Pelican Key, Roy Ozmer lived in self-seclusion as his cure for a lifelong battle with alcoholism. After a career as a journalist, actor and artist, Roy made the move in what he felt was a way to save his own life. For Roy, being a hermit was a cure for his alcoholism. His family would boat out to the key to visit him. Roy was clear that he withdrew from society because of his disease, not because he disliked people. He said, "I've foregone society, but if the world wants to come out and share a cup of coffee or talk over a problem, it's all right with me." He spent his remaining years writing poetry and drawing on the key.

Another example is Martha Frock, who lived in the Florida Everglades. She lived in a shack but described her living situation as "heaven." It's hard to imagine a swamp as home, but Martha picked her spot in the Everglades as her permanent home. She built what she described as a "shack" that rested on concrete blocks and thatched her roof with palm leaves. Using propane, she was able to have a stove, a fridge, a well pump and a few lights. Martha kept things simple and only ate a few times a week, surviving on what she could. It may have been a harsh existence, living side by side with the range of creatures that also call the Everglades home, but there isn't anywhere else Martha would have preferred to live.

This tradition of a solitary person living in self-isolation continued at Lily Spring with one of the last Floridians committed to this type of solitary lifestyle. Making his home on the banks of Lily Spring, Ed Watts lived there for decades. Starting in the mid-1980s, Ed made Lily Spring his permanent residence, even adding a mailbox that read "Wild Man Ed" to mark his home along the secluded spring. Ed's residency began over a period of several years, when he visited the spring occasionally after finding it accidentally after a long day paddling. Lily Spring became one of his favorite spots. His visits became overnight camping trips and then extended camping in the spring.

Since 1984, Ed Watts lived at Lily Spring. When Ed started living at the spring, it was filled with trash. Ed asked the owner of the spring if he could clean up the spring and, in exchange, swim in the spring naked. This was

"Naked Ed Looks to Redefine Place in World." From the *Gainesville* Sun.

a turning point for Ed Watts: he quickly became known to other paddlers visiting the spring as "Naked Ed." Eventually, he started to stay at the spring full time.

Part of the spring's appeal for Naked Ed was not just seclusion or a desire for a solitary life but also the spring water itself. The water at Lily Spring made Naked Ed feel remarkably better—reminiscent of claims about the healing properties of spots like White Springs. Living at the spring, a hard life for most, was easier than a life away from the healing water for Ed Watts.

Life in a spring made sense for Naked Ed. In his past lives, Ed operated a bar and worked as a commercial fisherman, a newspaper delivery man and a grocery store clerk. Naked Ed inherited a genetic condition called brittle bone disease, because of which he had broken many bones in his body, including his hips and fingers. His condition made his bones so fragile that he once broke a rib by sneezing. Broken fingers and toes were so common for Naked Ed that, at some point, he even stopped getting those checked by a doctor. Spending so much time swimming helped Naked Ed live with the condition.

The owner of Lily Spring gave Naked Ed a twenty-year lease that did not require any payment for living at the spring. The arrangement was that Naked Ed would be the spring's caretaker, and in exchange, it would be his home. The first version of Ed's spring-side home was an eight-by-ten shack with a palmetto leaf thatched roof. The second version was built on stilts to compensate for flooding. In later years, Ed added a telephone and electricity and built a deck, continuing to improve his spring-side home.

Somewhere along this journey, Naked Ed became a local legend. Canoe and kayakers who met Naked Ed at Lily Spring would paddle away and share their experience with others—unsurprisingly, given Ed's eccentric personality and lifestyle. He became so widely known that people would boat to Lily Spring in search of Naked Ed. Although he lived a hermit's lifestyle, Naked Ed liked talking to visitors. At times, he would put on a loincloth if it was a busy day on the river, but most of the time, he could be seen swimming nude in the crystal-clear waters of the spring.

Later, Naked Ed was referred to as "Ed the Oracle" or "Oracle of the Springs" due to his propensity for giving life advice to people who traveled into the spring to see him. His wisdom was captured on a variety of wooden signs that festooned the area. Some provided information about the spring; others shared life wisdom from Ed.

Life at the spring may have made sense for Ed, but he was not living a life of luxury. His home provided only rudimentary protection from the weather. And although the tree canopy did provide some protection from the sun, his skin was leathery from exposure. His body showed the wear of brittle bone disease, as his limbs were knobby from multiple breaks.

Despite what may seem like a hard life to many of us, Naked Ed claimed that his doctor credited the spring with saving his life. "That doctor thought the springs had a lot to do with me getting well," Ed said. "He said getting back here gave me an incentive to live." Caring for the spring and meeting the people who visited gave Naked Ed a purpose: "The canoers, they gave me something to do with my life. I have no clue what I'd be doing otherwise." Beyond the incentive to live, it's probable that the buoyancy provided by frequenting the spring water also provided some relief to Ed's medical condition.

Still, it's not as if Ed would have liked life another way, regardless of the medical benefits. In an interview with the *Gainesville Sun* about Lily Spring, Naked Ed said, "I'm part of the springs and they are part of me." He considered himself as much a part of the landscape as any of the trees in the forest surrounding the spring. Certainly, his work cleaning up the spring benefited the spring and the surrounding forest.

Life at the spring was not easy or idyllic, though. There was no protection from the many mosquitos and ticks found in this section of the river. Help was not nearby or easily accessed. For a time, Ed kept an old VW Rabbit at the gate where his mailbox sat, for trips to the grocery store or emergencies.

When visitors asked, Naked Ed insisted that he lived off forest creatures. He claimed he liked to eat insects and watched his visitors' reactions. When

The Oracle of Lily Spring, Naked Ed, shares his wisdom. *Michael Lusk.*

visitors asked him about his life and if he was afraid of animals like alligators attacking him, his reply was always, "The only animal man has to worry about is his fellow man."

In reality, Naked Ed went into town for groceries. He lived at the spring for a long time without electricity, although the newest version of his hut did have electricity and even a microwave. His collection of loincloths grew to twenty.

Some busy days on the river, Naked Ed had hundreds of visitors. He invited them to sit by his fire and trade stories. Although there were wooden signs warning visitors, to venture into Lily Spring was to accept that Ed's nudity might be part of the equation. Ed was known for being hospitable and giving his visitors popsicles.

Although Naked Ed's life along the banks of Lily Spring seemed tranquil, it was not without its share of drama. The second time Ed's home burned down, he alleged that it was intentional. Naked Ed claimed that this was done to try to get him to move out of the area. Lily Spring had been at the center of a long-running, complicated property dispute in which two different parties claimed they owned the land. Ed's presence at the spring

Naked Ed's welcome sign for Lily Spring. *Michael Lusk.*

became part of the dispute when one of the people claiming to own Lily Spring and the surrounding area went as far as to get a permit to demolish the shed, claiming it was a fire hazard. After the eventual fire (which Ed insisted was intentional), fans of Naked Ed offered to help rebuild. When asked, toward the end of his life, what the springs meant to him, he said, "I feel as attached to these springs as some people do to their families, now that I've been here all these years. I would love to be sitting out here as I draw my last breath."

Unfortunately for Naked Ed, that wish was not fulfilled. Ed Watts lost his battle with his health challenges and did not get to take his last breath at the spring. One can hope that Ed's spirit rests next to an eternal spring, but it is a certainty that Ed's presence will be felt at Lily Spring for a long time. Given the land disputes that were happening even while Naked Ed lived, the next chapter for Lily Spring remains uncertain.

Like White Spring and the legendary healing water that so many traveled to see, for Naked Ed, Lily Spring was that healing water. The spring water made a tangible difference in the quality of his life. The symbiotic relationship between Naked Ed and Lily Spring represents that communion with nature that so many visitors to the spring have felt. Naked Ed recognized that magical quality of the springs, just like what Bartram experienced at Manatee Spring. Ed devoted his life to that magic and was likely the last Florida hermit and the end of an era when Florida's wilderness offered those who wanted to live a different kind of life the chance to get close to nature and experience the sublime. Although the future of Lily Spring may be unknown, what is known for certain is that Lily Spring and Naked Ed are a legend that the surrounding communities will remember.

Chapter 11

GINNIE SPRINGS

Body Counts and a Whirling Vortex

There is no right or wrong way to enjoy the springs, but every spring is a different experience. When springhunting, a variety of options are out there waiting and ready. Ginnie Springs is more polarizing than most springs, however. Ginnie Springs inspires a "love it or leave it" reaction due to its vibe, very different than what other springs have to offer. Either Ginnie Springs is a fast favorite, or visitors leave unlikely to return again.

There is archaeological evidence that there was human activity at this spring (like at most Florida springs) for a long time. Due to the amount of flint and arrowheads found in the area, it is thought that the Timucua used the area to manufacture their tools and weapons. Ginnie Springs served as a weapons factory long before it was an entertainment destination.

The modern-day Ginnie Springs is named after Virginia, a woman who once washed her laundry at the spring. It is also famous as the location where Jacques Cousteau dove in 1974 and praised the crystal-clear water as "visibility forever." Prior to the arrival of Europeans, the Native Timucuan people established a settlement and stone works along the Santa Fe River nearby, where they manufactured stone tools. Divers still find artifacts from this settlement in the river, including broken pieces of flint, arrowheads and pottery.

The hydrology of Ginnie Springs is intricately linked to the Floridan aquifer, a vast underground reservoir that underlies much of Florida. Rainwater percolates through the ground, filtering through limestone layers, and eventually emerges at the springs, creating the clear, cool waters that Ginnie Springs is renowned for. The seven primary springs are Ginnie

Sign at Ginnie Springs. *Author's collection.*

Spring, Twin Spring, Dogwood Spring, Deer Spring, Devil's Eye Spring, Devil's Ear Spring and Little Devil Spring. Each of these springs is different from the others, and they all offer remarkable beauty.

The diverse ecosystem surrounding Ginnie Springs is teeming with aquatic life. The riverbanks and lush vegetation provide habitat for a wide variety of wildlife. Visitors may spot turtles, otters, manatees and numerous bird species, including herons, egrets and kingfishers. The water is very deep, much deeper than it first appears when you walk up to the spring. For less experienced swimmers, it is best to wear a life jacket, as there are few places shallow enough to stand.

Ginnie Springs is a system of seven crystal-clear freshwater springs, each with its unique characteristics. These springs are connected by a series of underwater tunnels and caverns, making them a popular destination for cave divers and snorkelers alike. The springs are known for their remarkable water clarity, which can extend up to two hundred feet in some areas. The Santa Fe River, which flows through the Ginnie Springs property, is ideal for tubing and canoeing. Visitors can rent tubes or bring their own and float down the picturesque river. There are tubes for sale and for rent at the general store at Ginnie Springs. Air pumps are available, also, if tubers bring their own tubes.

Three of the spring runs connect directly to the Santa Fe River. Tubing is allowed between each spring run, so tubers can start in one spring run and float out to the river. Once in the river, tubers can then navigate to the next spring run. If doing so with small children or less experienced swimmers, it is best to tube with a life jacket. The Santa Fe River may seem calm, but it does have currents that can be intense. Additionally, the Santa Fe River is not crystal-clear like the springs and the spring run. Although alligators usually cannot be bothered with the raucous spring visitors, it is worth showing the expected caution and acknowledging that the river is their home.

The springs' underwater caves and tunnels serve as critical habitats for cave-adapted species, including the Florida cave crayfish and the blind cave salamander. The delicate balance of this ecosystem underscores the importance of responsible exploration and conservation efforts. It is difficult to balance allowing the incredible volume of people that visit this spring each year access to do the activities they enjoy while also keeping the spring healthy and preventing problems like erosion on the banks of the springs.

Part of what draws so many crowds to this spring is that the underwater caves and caverns are a draw for cave divers, scuba divers and snorkelers. Devil's Eye and Devil's Ear are renowned dive sites, offering breathtaking underwater experiences. These spots are not just renowned in Florida but are also a top spot internationally for scuba diving and cave diving. The park also offers onsite rentals for equipment, making it a convenient spot for divers to easily get in the water with everything needed for a successful dive.

The spring did not become an internationally recognized hot spot for scuba until the last sixty years. This spring's history is also linked directly to the development of scuba diving as a sport. The idea of a secret spring swept through scuba circles, and the spring became a favorite among members of this new sport. Although the sheer beauty of the spring has always been a draw, other factors were temperature and location. Although Floridians might consider the spring's constant temperature of seventy-two degrees cold, divers from other locations found the water inviting, even in winter, since they were used to much colder temperatures. The location was also a factor in the early popularity of Ginnie Springs because scuba divers would stop there on their return trips from the Florida Keys. When it was already a popular spot, Jacques Cousteau stated that the water was the clearest he had ever experienced and described the spring as "visibility forever," which supercharged the dive community's interest even more. Divers wanted to see this water thought to be the clearest in the world.

Ginnie Springs. *Wikimedia Commons.*

Several springs like Silver Springs or Weeki Wachee owe their renown to well-planned and well-executed advertising campaigns. Without a single advertisement, Ginnie Springs became a sensation among diving enthusiasts. Even now, when it draws more crowds each weekend than most springs, Ginnie Springs does not advertise extensively. Much of its popularity is due to word of mouth and social media influencers who document the wilder side of the spring.

Ginnie Springs does offer both picnic tables and camping, but this is not your typical "commune with nature" camping or picnic experience. Ginnie Springs is a natural wonder and an epic party rolled into one. It's like walking into a big college party. Especially during the on-season and on weekends, it can be hard to even see the water for all the tubes, rafts, paddleboards and kayaks that fill it. There is music playing, usually for more than one party, from first thing in the morning until late at night. Although the park does state that there are quiet hours, that is more of a suggestion than a hard-and-fast rule.

Ginnie Springs is a bit different. Undeniably, under the water, it is one of the most incredible springs to see. Some of the springs feel sacred or spiritual, but Ginnie Springs has a completely different vibe. Above the water at Ginnie Springs is on a whole other spectrum. With fewer restrictions since the spring is privately owned, walking up to the spring feels like a large, communal, endless summer party. Although the park staff tries to clean up after the wild raging parties, there are bottles and cans everywhere, on weekends especially.

There are areas with less crowds, such as the smaller springs, which almost have a feeling of intimacy. Camping near the springs and seeing them at night by only starlight makes for a moment long remembered, but keep in mind that the serenity of the springs is a backdrop to an ongoing party, like the drug-fueled parties at Ichetucknee Spring in decades past.

When the Wray family purchased the land and Ginnie Springs, it was initially kept private. However, divers would walk through the woods to explore the spring anyway. Later, the Wray family would develop the land as the party destination it is now.

The campgrounds continue the summer party feel with a higher level of intensity. The main campground feels like a fraternity block party on weekends but with a wider variety of participants. There are visitors from all over the state who come for the raging party—reducing the spring to a nice addition to the party. There are hammocks along the water used to recover from hangovers. The primitive camping area is even less regulated than the main campground: it's an "anything goes" environment. There are local legends about everything from mass, drug-fueled orgies and group psychedelic drug experiences to fringe groups that want to gaze at the stars while practicing their own spirituality. Ginnie Springs is the modern-day version of the party spot that was Ichetucknee Spring in the last century.

The party atmosphere, however, is not the whole story. Ginnie Springs is a remarkably beautiful spring, so whether or not visitors want to partake in the wildness happening in the woods, the spring is worth seeing for its unique and stunning beauty.

It is the park being privately owned that makes the environment at this spring so different. Since the 1970s, Ginnie Springs has been owned by the Wray family. Before the family purchased the property, the popularity of the spot had already taken off among the dive community. As diving grew in popularity, so did the number of deaths associated with it in Florida. As a reaction to the deaths, for a time, Florida parks disallowed—or in some spots stopped—access to divers. This made private spots like Ginnie Springs even more popular in the community.

When the current owner purchased the spring, he described it as: "There was nothing but a little dirt road here. It was a mess." Certainly the spring would not have been easily accessible to the public. There was clear evidence, however, that even if the spring was not easily accessible, the public was making it to the spring. Wray went on to describe his first visit to the spring, saying, "There was trash everywhere. It did not look inviting."

Even before purchasing the property, Wray would camp on the land for extended periods of time. Not yet a scuba diver himself, he talked to trespassing divers and realized that they would always come to the spot, no matter what. There was no stopping the influx of divers who wanted to see Ginnie Springs for themselves.

Wray was in an interesting position to acquire Ginnie Springs and traded the current owner an island in Pinellas coast for the spring and the property around it. Wray's initial vision was to allow access to the spring while keeping it as close to the feeling of seclusion and privacy that you get when you visit the spring alone—in other words, minimal development.

When the Wray family first purchased the property, they built a home, intended as a quiet spot. However, nothing they did or tried kept people from trespassing and entering the spring. Despite attempts to limit access, divers would still get into the cave system, and twenty-two divers died in Ginnie Springs in the second half of the twentieth century. That number represents the known deaths, whereas the actual number could be much higher if divers entered the water without anyone knowing. The danger of the dives could have also been part of the appeal. The dive seems easier than it is. The vast cave systems are intriguing, and in the early days of scuba, there was less regulation regarding who could dive and to what extent training was required or even needed. As appealing as Ginnie Springs was to divers, it was also a dangerous place.

The Wrays' next strategy to keep divers out of the cave system was to install an iron grate over the most dangerous section of the caves. The family also put up signs warning divers that the area was dangerous. According to *Divers Direct*, Florida holds the highest number of cave-diving deaths, and these springs hold the highest number of deaths in a spring. The interconnected caves between the springs are part of the appeal. As the Wray family tried to keep the spring private, divers continued to show up to use the spring anyway. The family eventually opened the site to the public once again.

Whether it is because of the frequent use of mind-altering substances by visitors to Ginnie Springs or the sheer number of people who have lost their lives in its waters, there are some legends centered on this spring. The remote nature of the park also leaves room for some strange stories to emerge.

Like Manatee Spring and Weeki Wachee, Ginnie Springs was also rumored to have mermaids. Believers state that the mermaids would swim into the spring via the Santa Fe River and that in the moonlight, the mermaids could be seen swimming through the caves and caverns of Ginnie Springs. Given the deep and extensive nature of the cave system, it holds an element of mystery that makes some believe mermaids are a possibility. Florida mermaids, or people cosplaying as mermaids in full costume, are definitely present at Ginnie Springs most weekends.

A second legend is also attributable to the remote nature of Ginnie Springs. Although Ginnie Springs is relatively close to High Springs, the area around the park is rural, and the park itself sits on six hundred acres (about the area of Central Park in New York City). This next tale taps into the Skunk Ape legend that is popular in Florida.

The Skunk Ape is also called the Skunk Cabbage Man and other similar names due to the smell of the creature, which is described pungent swamp water. There is an official Skunk Ape Headquarters in Ochopee, Florida, that collects stories and sightings of the creature. It all began with an old video clip showing a cryptid in a Florida swamp. At Ginnie Springs, the cryptid tales get a little more sinister, claiming that the creature has purple skin and long, sharp claws. Sightings include tales of swimmers having their legs grabbed by the creature's slimy hands, its claws leaving deep gashes. Most of the sightings seem to place the purple Skunk Ape in the forests surrounding Ginnie Springs. Visitors to the primitive camping area report seeing all sorts of creatures, with the Skunk Ape being the most popular. Some suggest that the creature at Ginnie Springs is the wolflike "Long Ears" described in Seminole legends. Others claim to have smelled the ape rather than seen it, stating that the smell is like a skunk's. One reason for the smell could be the amount of marijuana that is consumed at the spring, however.

The Skunk Ape may visit the forests of Ginnie Springs, but an even more frequent and persistent visitor is the Grim Reaper. The allure of cave diving at the spring attracts seasoned professionals as well as untrained divers. Cave diving is already dangerous, but Ginnie Springs offers twenty-five thousand feet of mapped caves that are irresistible. Although the reported number of deaths at Ginnie Springs varies depending on the source, some estimates suggest that 150 people have died at Ginnie Springs since 1950. That is only one of the ways to die at Ginnie Springs. Recent deaths include a car accident in the parking lot, an accidental shootings, a murder at the campground and a recent multi-fatality shooting.

Divers who have survived exploration trips through Ginnie Springs have reported feeling a vortex under the water that swirls them around and disorients them. This phenomenon has been reported often enough that experienced divers warn new divers that they might feel the vortex and tell them how they can recover if it happens. The vortex could be a product of the large volume of water that moves through the springhead or a simple current in the water. Legend has it that the vortex is the souls of the deceased divers trying to find their way out of the spring. Some of these divers' bodies were never recovered from the cave depths.

Yet another legend claims there is a ghostly, luminescent canoe that floats in from the rivers and down each of the spring runs before returning to the river once again. Like at many of the other springs, there are also reports of strange noises in the forest that don't have an apparent cause. While a guarantee cannot be made that visitors will see a supernatural canoe or see the Skunk Ape, it is certain that, at Ginnie Springs, strange noises in the woods will be heard.

On a more positive note, the management of Ginnie Springs has demonstrated a commitment to environmental conservation and preservation. Efforts have been made to minimize the impact of visitors on the delicate ecosystem. Regular water quality testing is conducted to ensure the springs' water remains pristine and safe for recreation. Ginnie Springs actively promotes responsible outdoor ethics and environmental stewardship among visitors, including guidelines for cave diving and snorkeling on best practices to avoid environmental damage. The owners have undertaken projects to restore and protect the natural habitats around the springs.

Despite its natural beauty and conservation efforts, Ginnie Springs has not been without its share of challenges and controversies. Concerns have been raised about the increasing number of visitors and the potential impacts on the fragile ecosystem. Balancing the desire for access with the need for preservation remains an ongoing challenge for the management and conservationists.

Additionally, debates have arisen over issues such as water extraction permits and potential impacts on groundwater levels. These challenges underscore the importance of ongoing dialogue and responsible management to ensure the long-term sustainability of this natural wonder.

Ginnie Springs, Florida, stands as a testament to the beauty and wonder of the natural world. Its crystal-clear springs, diverse ecosystem and rich human history make it a cherished destination for outdoor enthusiasts and nature lovers. The beauty of these springs cannot be overstated, but the atmosphere is much different than that of most other springs. Weekends at Ginnie Springs are extremely busy, and the behavior of the guests can be unpredictable and dangerous. Families with less experienced swimmers should avoid this spring during holiday weekends because the crowds make swimming and navigating the spring difficult. For a more laid-back experience, try visiting this spring during a weekday in the cooler months. Due to a shooting over Memorial Day weekend in 2024, however, visit this spring with extreme caution.

The power of suggestion and the atmosphere of a place can contribute to paranormal experiences. When visitors arrive at Ginnie Springs with the expectation of having a wild time and possibly encountering the supernatural, even ordinary events can take on an eerie, supernatural quality. As for the paranormal stories that shroud Ginnie Springs, they add an intriguing layer of mystique to the already captivating location. Whether you are a believer in the supernatural or a staunch skeptic, the tales of ghostly canoes, mournful cries and shadowy figures will continue to spark curiosity and wonder among those who venture into this remarkable natural haven.

In the end, Ginnie Springs invites visitors to embrace both the natural world and the mysteries that may lie just beyond our understanding. Whether a trip to Ginnie Springs is to seek adventure or a chance to connect with the unknown, this is a place where the line between the ordinary and the extraordinary blurs. The Memorial Day shootings, in addition to all the other lives this spring has claimed, darken the wonder and appeal of this majestic place.

BIBLIOGRAPHY

Anderson, Charles H. "Way Down Upon the Suwannee." *Florida Wildlife* 3 (February 1950): 11, 14.

BackpackerVerse. "Swimming Holes in Florida." https://backpackerverse.com/swimming-holes-in-florida.

Behnke, Patricia. "Old Timers Remember—Ichetucknee Springs." Florida Department of Environmental Protection. Accessed January 26, 2024. https://www.ichetuckneealliance.org.

Beloved Blue River. "Historical Timeline." https://belovedblueriver.org/the-human-role/history-and-long-term-trends/historical-timeline.

Burt, Al. "Allure of Springs." *Gainesville Sun*, April 20, 2003.

Caldwell, John M. "Steamboat Madison." Unpublished manuscript, copied by Historical Records Survey, State Archives Survey, 1937.

Coleman, Jenn, and Ed Coleman. "True Scary Stories of the Suwannee." *Coleman Concierge.* https://www.colemanconcierge.com/true-scary-stories-suwannee.

Columbus, Christopher. "Journal of the First Voyage of Columbus." In *Journal of Christopher Columbus (During His First Voyage, 1492–93), and Documents Relating to the Voyages of John Cabot and Gaspar Corte Real.* Edited and translated by Clements R. Markham. London: Hakluyt Society, 1893. 224–25.

Crabbe, Nathan. "Water Wars Put Springs in Spotlight." *Gainesville Sun*, May 14, 2007. https://www.gainesville.com/story/news/2007/05/14/water-wars-put-springs-in-spotlight/31524484007.

Desolation Florida. "The Telford Hotel of White Springs." September 2016. http://www.desolationflorida.com/2016/09/the-telford-hotel-of-white-springs.html.

Divers Direct. "Florida's Spookiest Dive Sites." https://www.diversdirect.com/w/floridas-spookiest-dive-sites.

Florida Museum of Natural History. "The Town of the White King." Exhibit sign, December 10, 2023.

Florida State Parks. "Ichetucknee Springs State Park." https://www.floridastateparks.org/parks-and-trails/ichetucknee-springs-state-park.

———. "Wreck of the Madison." https://www.floridastateparks.org/learn/wreck-madison.

Gainesville Creeks. "Boulware Springs." https://www.gainesvillecreeks.org/cw-boulware-springs.

Gallagher, Peter B. "Florida's Underground Frontier." *St. Petersburg Times*, November 29, 1983.

Gilmore, Tim. "Tall Tales of Giant Timucua or the Myth of Indigenous Giants." Jacksonville Psychogeography, June 22, 2022. https://jaxpsychogeo.com/all-over-town/tall-tales-of-giant-timucua-or-the-myth-of-indigenous-giants.

Guide to Greater Gainesville. "The Good Ole Days at the Grady House." https://guidetogreatergainesville.com/the-good-ole-days-at-the-grady-house.

Haunted Places. "Grady House Bed and Breakfast." https://www.hauntedplaces.org/item/grady-house-bed-and-breakfast.

Hermitary. "Florida Everglades Hermits, 1940s to 1980." https://www.hermitary.com/articles/everglades.html.

Historic White Springs. "History." https://www.historicwhitesprings.com/history-1.

Ichetucknee Springs Park. "Ichetucknee Springs Park History." https://ichetuckneesprings.com/park-history.

King, Roger (interviewee), and Marna Weston (interviewer). "Interview with Roger King 2011-06-30." Samuel Proctor Oral History Program, 2011. https://original-ufdc.uflib.ufl.edu/AA00084888/00001.

Kirkpatrick, Mary Alice. "William Bartram's Legacy." Documenting the American South. https://docsouth.unc.edu/highlights/bartram.html.

Klinkenberg, Jeff. "The Wild Man of Lilly Spring." *Tampa Bay Times*, August 13, 2000. Updated September 27, 2005. https://www.tampabay.com/archive/2000/08/13/the-wild-man-of-lilly-spring.

Milanich, Jerald T. "Archaeological Evidence of Colonialism: Franciscan Spanish Missions in La Florida." *Missionalia* 32, no. 3 (November 2004): 332–56. https://journals.co.za/doi/pdf/10.10520/AJA02569507_1048.

Newcomer, Daniel. "The Wreck of the Steamship Madison." Clio, December 12, 2015. https://theclio.com/entry/20802.

On the Road in Florida with Idelle (blog). "High Springs Museum." November 6, 2018. https://ontheroadinfloridawithidelle.wordpress.com/2018/11/06/high-springs-museum.

Pittman, Craig. "Ginnie Springs Owner Fights Off Threats." *Tampa Bay Times*. https://www.tampabay.com.

———. "The Ripple Effect: Florida Springs." *Flamingo Magazine*, May 27, 2016. https://flamingomag.com/2016/05/27/the-ripple-effect-florida-springs.

Rosborough, Bob. "LSD: Movement, Religion, Obsession." *Florida Alligator*, October 13, 1966. Via University of Florida Digital Collections. https://ufdc.ufl.edu/UF00028291/02556/images.

Springs Eternal Project. "History: Ichetucknee Springs." https://springseternalproject.org/springs/ichetucknee/history-itchetucknee.

St. Petersburg Times. "Ginnie Springs." November 29, 1983.

Top Hot Springs. "Fanning Springs State Park History." https://www.tophotsprings.com/fanning-springs-state-park-florida.

Warren, Michael. "Manatee Springs State Park." Florida Traveler. https://floridatraveler.com/manatee-springs-state-park.

WCJB. "Home of High Springs' Naked Ed Burns Down." February 1, 2022. https://www.wcjb.com/2022/02/01/home-high-springs-naked-ed-burns-down.

Wheeler, Ryan J., and Christine L. Newman. "An Assessment of Cultural Resources at the Troy Springs Conservation and Recreation Area." Bureau of Archaeological Research, C.A.R.L. Archaeological Survey, September 1996. http://files.usgwarchives.net/fl/lafayette/history/carltroy.txt.

Works Progress Administration, Historical Records Survey. *History of Lafayette County.* September 1940. State Archives of Florida, Florida Memory. https://www.floridamemory.com/items/show/321123.

ABOUT THE AUTHOR

Holly Sprinkle is a writer and instructional designer in Gainesville, Florida. She has a master of arts in professional writing from Western Carolina University. After starting her career as a college professor in writing and literature classrooms, Holly embarked on a career as writer, editor and online course developer. Ms. Sprinkle made her mark on the early online teaching world by advocating for standards including high visual appeal and maximizing design principles to improve knowledge transfer. Holly enjoys spending time with her three boys exploring and photographing Florida's natural wonders. Their favorite spots include Florida's springs, beaches and hiking trails.